College Mathematics II

Personal Study Notes

Special Functions
Fourier Transforms
Laplace Transforms

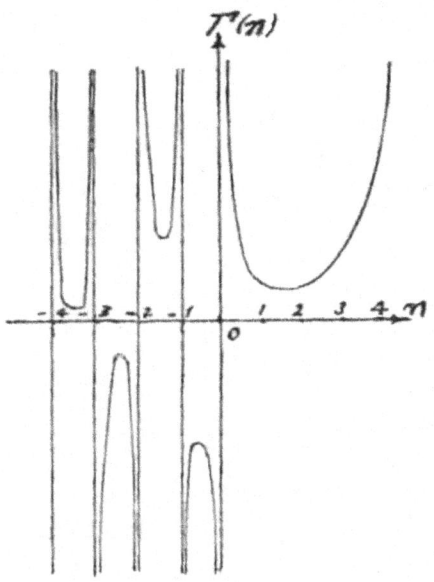

By Mohamed F. El-Hewie

TABLE OF CONTENTS

CHAPTER 7

SPECIAL FUNCTIONS

19121 72

Special Functions

Gamma and Beta Functions

1. Gamma Functions

1. Gamma f^n

The 2nd Eulerian integral

Usually denoted by $\Gamma(n)$, is defined as follows

$$\Gamma(n) = \int_0^\infty e^{-x} x^{n-1} \, dx \qquad n > 0$$

The condition of $n > 0$ is necessary for the convergence of the integral.

$$\Gamma(1) = \int_0^\infty e^{-x} \, dx = -\left[e^{-x}\right]_0^\infty = 1$$

$$\Gamma(n+1) = \int_0^\infty e^{-x} x^n \, dx = -\int_0^\infty x^n \, d e^{-x}$$

$$= -\left[x^n e^{-x}\right]_0^\infty + \int_0^\infty e^{-x} \, dx^n$$

$$= n \int_0^\infty e^{-x} x^{n-1} \, dx \qquad \because x^n = z$$
$$n x^{n-1} dx = dx^n$$

$$= n \, \Gamma(n)$$

1.1. Properties of Gamma functions

$$\boxed{\Gamma(n+1) = n \, \Gamma(n)}$$

Now suppose that n is a positive integer

$$\Gamma(n) = (n-1) \, \Gamma(n-1)$$
$$= (n-1)(n-2)(n-3) \, \Gamma(n-3)$$

$$= (n-1)!$$

$$\boxed{\Gamma(n) = (n-1)!} \qquad \text{if } n \text{ is +ve integer.}$$

e.g.
$$\Gamma(6) = 5!$$
$$\Gamma(5) = 4!$$

Thus $\Gamma(n)$ may be considered as extension of factorials

$$\Gamma\left(\tfrac{1}{2}\right) = \int_0^\infty e^{-x} x^{-\frac{1}{2}} \, dx$$

6

Put $x = y^2$; $dx = 2y\, dy$

$$\Gamma\left(\tfrac{1}{2}\right) = \int_0^\infty e^{-y^2} y^{-1} \cdot 2y\, dy$$

$$= \left(2\int_0^\infty e^{-y^2} dy\right) = 2\frac{\sqrt{\pi}}{2} = \sqrt{\pi}$$

$$\boxed{\Gamma\left(\tfrac{1}{2}\right) = \sqrt{\pi}}$$

For negative values of n, $\Gamma(n)$ is defined as followes

$$\Gamma(n) = \frac{\Gamma(n+1)}{n}$$

eg $\quad \Gamma\left(-\tfrac{1}{2}\right) = \dfrac{\Gamma\left(\tfrac{1}{2}\right)}{-\tfrac{1}{2}} = -2\sqrt{\pi}$

$\Gamma\left(-\tfrac{3}{2}\right) = \dfrac{\Gamma\left(-\tfrac{1}{2}\right)}{-\tfrac{3}{2}} = \dfrac{4}{3}\sqrt{\pi}$

$\Gamma\left(-\tfrac{5}{2}\right) = \dfrac{\Gamma\left(-\tfrac{3}{2}\right)}{-\tfrac{5}{2}} = -\dfrac{8}{15}\sqrt{\pi}$

Signs are repeated $-, +, -, +, -$ and coeff^ts of $\sqrt{\pi}$
decrease $\quad -2, \tfrac{4}{3}, -\tfrac{8}{15}, +, -, \ldots$

1.2. Gamma function graph

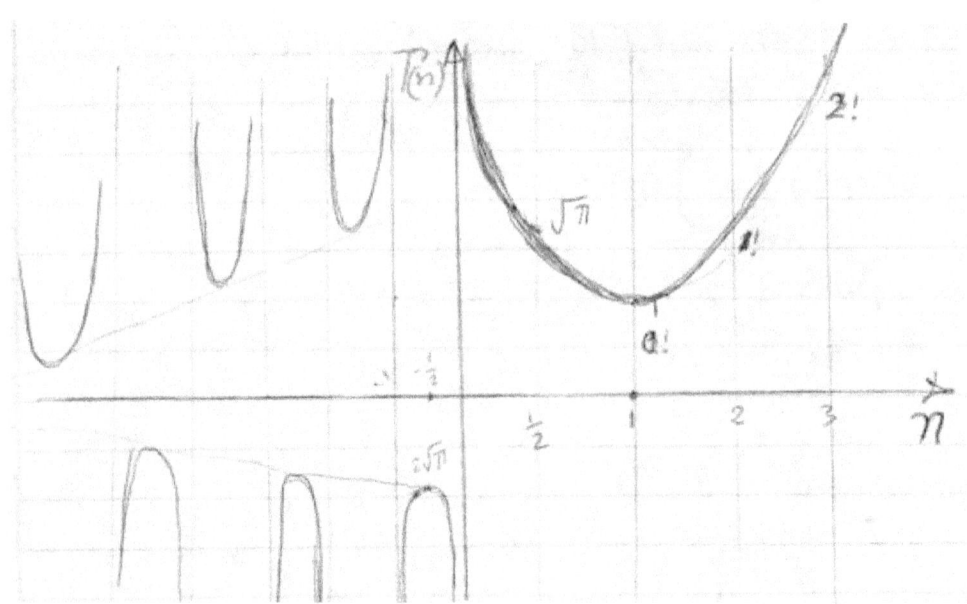

$\Gamma(n)$ is not defined at $n = 0$ & negative integral values of n

2. Beta Functions

[2]. <u>Beta f^n</u> :- The 1st Eulerian integral usually denoted by $B(m,n)$ is defined as follows:-

$$B(m,n) = \int_0^1 x^{m-1}(1-x)^{n-1} dx$$

$m > 0$
$n > 0$

case $B(1,1) = \int_0^1 dx = 1$

m, n are interchangeable جنس ريد ك

for $x = 1 - y$
$dx = -dy$
$x = 0 \qquad y = 1$
$x = 1 \qquad y = 0$

8

$$B(m,n) = \int_1^0 - (1-y)^{m-1} (y)^{n-1} \, dy$$

$$\beta(m,n) = \int_0^1 (y)^{n-1} (1-y)^{m-1} \, dy$$

In the expansion
of $\beta(m,n)$ put $x = \sin^2\theta$,

$$dx = 2\sin\theta\cos\theta \, d\theta$$

$$x = 0 \qquad \theta = 0 \qquad x = 1 \qquad \theta = \frac{\pi}{2}$$

$$\beta(m,n) = \int_0^{\frac{\pi}{2}} \sin^{2m-2}\theta \cdot \cos^{2n-2}\theta \cdot 2\sin\theta\cos\theta \, d\theta$$

$$= 2\int_0^{\frac{\pi}{2}} \sin^{2m-1}\theta \cdot \cos^{2n-1}\theta \cdot d\theta$$

2.1 Relation between Beta and Gamma functions

We shall now prove the following relation:

$$\therefore \quad \beta(m,n) = \frac{\Gamma(m) \cdot \Gamma(n)}{\Gamma(m+n)}$$

9

The proof of Fundamental relation connection.

$\beta \& \Gamma$ fⁿ

$$\beta(m,n) = \frac{\Gamma(m) \cdot \Gamma(n)}{\Gamma(m+n)}$$

We shall now show that.

$$\Gamma(n) = \int_0^\infty e^{-x} \cdot x^{n-1} \, dx \qquad \text{put } x = y^2$$
$$dx = 2y\,dy$$

$$= \int_0^\infty e^{-y^2} y^{2n-2} \cdot 2y\,dy$$

$$= 2\int_0^\infty e^{-y^2} \cdot y^{2n-1} \, dy \qquad \text{————①}$$

similarily.

$$\Gamma(m) = 2\int_0^\infty e^{-x^2} \cdot x^{2m-1} \, dx \qquad \text{————②}$$

$$\therefore \Gamma(n)\,\Gamma(m) = 4\int_0^\infty \int_0^\infty e^{-(y^2+x^2)} y^{2n-1} \cdot x^{2m-1} \, dy\,dx$$

$$\text{————③}$$

Transform into polar coordinates
$$x = r\cos\theta$$
$$y = r\sin\theta$$

$$x = r\cos\theta \qquad\qquad dx = -r\sin\theta\, d\theta + \cos\theta\, dr$$
$$y = r\sin\theta \qquad\qquad dy = r\cos\theta\, d\theta + \sin\theta\, dr$$

$$dy\, dx = -r^2\sin\theta\cos\theta\, \overline{d\theta}^2 + r\cos^2\theta\, dr\, d\theta$$
$$+ \cos\theta\sin\theta\, \overline{dr}^2 - r\sin^2\theta\, dr\, d\theta$$

$$= \cos\theta\sin\theta\left(-r^2\overline{d\theta}^2 + \overline{dr}^2\right) + r(1 - \sin^2\theta)\, dr\, d\theta$$
$$- r\sin^2\theta\, dr\, d\theta$$

$$= \cos\theta\sin\theta\left(\overline{dr}^2 - r^2\overline{d\theta}^2\right) + r\, dr\, d\theta -$$
$$\underset{\text{2nd order}}{\qquad} r\sin^2\theta\, d\theta\, dr\, (-1\, -1)$$

$$dx\, dy = r\, d\theta\, dr$$

$$\Gamma_{(n)}\,\Gamma_{(m)} = 4 \int_0^{\frac{\pi}{2}} \int_0^{\infty} e^{-r^2}\, r^{2m-1}\, \cos^{2m-1}\theta\; r^{2n-1}\, \sin^{2n-1}\theta\; r\, d\theta\cdot dr$$

$$\therefore\quad dx\, dy = r\, d\theta\cdot dr$$

$$y, x \longrightarrow 0 \qquad \theta = 0 \qquad\quad y\longleftarrow 0$$
$$y, x \longrightarrow \infty \qquad \theta = \frac{\pi}{2} \qquad r = \infty$$

$$\Gamma_{(n)}\,\Gamma_{(m)} = 2\int_0^{\frac{\pi}{2}} \cos^{2m-1}\theta\, \sin^{2n-1}\theta\, d\theta \cdot \frac{2}{2} \int_0^{\infty} e^{-r^2}\, r^{2m+2n^2}\, \overline{dr}^2$$

$$= \beta(m, n)\cdot \Gamma(m+n)$$

$$\therefore \quad \beta(m,n) = \frac{\Gamma(m) \cdot \Gamma(n)}{\Gamma(m+n)}$$

2.2. Applications of Beta function or integral

Applications on this relation :- the evaluation of integrals of the type :-

$$\int_0^{\frac{\pi}{2}} \sin^p \theta \cos^q \theta \, d\theta$$

Important integral from the application veiw. we have already seen that.

$$\beta(m,n) = 2 \int_0^{\frac{\pi}{2}} \sin^{2m-1} \theta \cos^{2n-1} \theta \, d\theta$$

Put
$$2m-1 = p \qquad \qquad \therefore m = \frac{p+1}{2}$$
$$2n-1 = q \qquad \qquad n = \frac{q+1}{2}$$

$$\therefore \int_0^{\frac{\pi}{2}} \sin^p \theta \cos^q \theta \, d\theta = \frac{1}{2} \beta\left(\frac{p+1}{2}, \frac{q+1}{2}\right)$$

$$= \frac{1}{2} \frac{\Gamma\left(\frac{p+1}{2}\right) \Gamma\left(\frac{q+1}{2}\right)}{\Gamma\left(\frac{p+q}{2} + 1\right)}$$

e.g

ex. $\int_0^{\frac{\pi}{2}} \sin^4\theta \cos^6\theta \, d\theta = \frac{1}{2}\beta\left(\frac{5}{2}, \frac{7}{2}\right)$

$$= \frac{1}{2} \frac{\Gamma\left(\frac{5}{2}\right)\Gamma\left(\frac{7}{2}\right)}{\Gamma\left(\frac{10}{2}+1\right)}$$

$$= \frac{1}{2} \frac{\left(\frac{3}{2} \times \frac{1}{2}\sqrt{\pi} \cdot \frac{5}{2} \times \frac{3}{2} \times \frac{1}{2}\sqrt{\pi}\right)}{5!}$$

ex: $\int_0^{\frac{\pi}{2}} \sin^8\theta \, d\theta$

$$= \int_0^{\frac{\pi}{2}} \sin^8\theta \underline{\cos^0\theta} \, d\theta$$

$$= \frac{1}{2}\beta\left(\frac{9}{2}, \frac{1}{2}\right)$$

$$= \frac{1}{2} \frac{\Gamma\left(\frac{9}{2}\right) \cdot \Gamma\left(\frac{1}{2}\right)}{\Gamma(5)}$$

$$= \frac{1}{2} \frac{\left(\frac{7}{2} \times \frac{5}{2} \times \frac{3}{2} \times \frac{1}{2}\sqrt{\pi} \cdot \sqrt{\pi}\right)}{4!}$$

13

again

$$\beta(m,n) = \frac{\Gamma(m)\cdot\Gamma(n)}{\Gamma(m+n)}$$

$$\Gamma(m) = \int_0^\infty e^{-x} x^{m-1} \, dx \qquad \text{put } x = y^2$$
$$\qquad\qquad\qquad\qquad\qquad\qquad\qquad dx = 2y\,dy$$

$$= 2\int_0^\infty e^{-y^2} y^{2m-2} \cdot y \, dy$$

$$= 2\int_0^\infty e^{-y^2} y^{2m-1} \, dy \qquad\qquad ①$$

$$\Gamma(n) = 2\int_0^\infty e^{-x^2} x^{2n-1} \, dx \qquad\qquad ②$$

$$\Gamma(m)\Gamma(n) = 4\int_0^\infty \int_0^\infty e^{-(x^2+y^2)} y^{2m-1} x^{2n-1} \, dy\,dx \qquad ③$$

$$= 4\int_0^{\frac{\pi}{2}} \int_0^\infty e^{-r^2} r^{2m-1} \sin\theta^{2m-1} r^{2n-1} \cos\theta^{2n-1} r\,d\theta\,dr$$

$$= 4\int_0^{\frac{\pi}{2}} \sin^{2m-2} \cdot \sin\theta \cdot \cos^{2n-2} \cdot \cos\theta \, d\theta \cdot \int_0^\infty$$

$$\int_0^\infty e^{-r^2} r^{2m+2n-2} \, r\,dr$$

$$= \frac{2}{2}\int_0^{\frac{\pi}{2}} \sin^{2m-2} \cos^{2n-2} \, d\sin^2\theta \cdot \int_0^\infty$$

$$\frac{2}{2}\int_0^\infty e^{-r^2} r^{2m+2n-2} \, dr^2$$

14

$$= \int_0^{\frac{\pi}{2}} \sin^{2(m-1)}\theta \cdot (1-\sin^2\theta)^{(n-1)} \, d\sin^2\theta \cdot$$

$$\int_0^{\infty} e^{-\mu^2} \cdot \mu^{2(m+n-1)} \, d\mu^2$$

Put $\sin^2\theta = Y$

$\& \ \mu^2 = Z$

$$\therefore \Gamma(m) \cdot \Gamma(n) = \int_0^1 Y^{m-1} \cdot (1-Y)^{n-1} \, dY \cdot$$

$$\int_0^{\infty} e^{-Z} \cdot Z^{m+n-1} \, dZ$$

$\theta = 0$	$Y = 0$
$\theta = \frac{\pi}{2}$	$Y = 1$
$\mu = 0$	$Z = 0$
$\mu = \infty$	$Z = \infty$

$$= \beta(m,n) \cdot \Gamma(m+n)$$

$$\therefore \beta(m,n) = \frac{\Gamma(m) \ \Gamma(n)}{\Gamma(m+n)}$$

$$\boxed{\int_0^{\infty} e^{-y^2} \, dy = \frac{\sqrt{\pi}}{2}}$$

Put $I = \int_0^{\infty} e^{-y^2} \, dy$

$$\therefore I^2 = \int_0^{\infty} \int_0^{\infty} e^{-y^2} \cdot e^{-x^2} \, dy \, dx$$

$r \, d\theta \, dr = dx \, dy$

$$= \int_0^\infty \int_0^\infty e^{-(x^2+y^2)} \, dx\,dy$$

$$= \int_0^{\frac{\pi}{2}} \int_0^\infty e^{-r^2} \, r\,d\theta\,dr.$$

$$\therefore I^2 = \frac{1}{2} \int_0^\infty \left[\theta e^{-r^2} \right]_0^{\frac{\pi}{2}} dr^2 = \frac{1}{2} \int_0^\infty \frac{\pi}{2} e^{-r^2} dr^2$$

$$= -\frac{\pi}{4} \left[e^{-r^2} \right]_0^\infty = -\frac{\pi}{4} \left[0 - 1 \right] = \frac{\pi}{4}$$

$$\therefore I = \frac{\sqrt{\pi}}{2} = \int_0^\infty e^{-y^2} dy$$

$$\boxed{\Gamma\!\left(\frac{1}{2}\right) = \sqrt{\pi}}$$

$$\Gamma\!\left(\frac{1}{2}\right) = \int_0^\infty e^{-x} x^{\frac{1}{2}-1} \, dx$$

$$= \int_0^\infty e^{-x} x^{-\frac{1}{2}} \, dx \qquad \text{put } x = y^2$$
$$\underline{\hspace{2cm}} dx = 2y\,dy$$

$$= 2 \int_0^\infty e^{-y^2} y^{-1} \, y\,dy$$

$$= 2 \int_0^\infty e^{-y^2} \, dy$$

$$= 2 \, \frac{\sqrt{\pi}}{2} = \sqrt{\pi}$$

3. Bessel Functions

These functions are originally due to the German astronomer Bessel who used them in his researches on the motion of planets in the first quarter of the nineteenth century. These functions are of fundamental importance since they are applied to a large number of problems which are of interest to engineers.

26\2\72

3/ **Bessel Functions** (German astronomer).

These f^{ns} are discovered after the german astronomer Bessel who investigated the motion of planets there f^{ns} are sol^{ns} of Bessel differential eq^{ns} of order (n) given by

$$x^2 \frac{d^2y}{dx^2} + x \frac{dy}{dx} + (x^2 - n^2)y = 0 \quad \text{———} \quad (1)$$

Here word "order" is used in another mean that the order here mean the value of (n).

Applications:—
1- Planetary motion.
2. Vibrations of a circular membranes.
3. Oscilations of a circular sheet of water
4- High frequency currents in cylindrical conductors

17

5- Problems of conduction field of heat.

6. Oscilations of a flixable chains, & hanging freely under gravity.

In most practicly problems we meat eq^n ① in the

$$x^2 \frac{d^2y}{dx^2} + x \frac{dy}{dx} + (k^2x^2 - n^2)\, y = 0 \quad —② $$

To reduce eq^n (2) to the from ① put $kx = t$

i) $kx = t \qquad \frac{dt}{dx} = k$

ii) $\frac{dy}{dx} = \frac{dy}{dt}\frac{dt}{dx} = k\frac{dy}{dt}$

iii) $\frac{d^2y}{dx^2} = k\frac{d}{dx}\left(\frac{dy}{dt}\right) = k\frac{d}{dt}\left(\frac{dy}{dt}\right)\frac{dt}{dx} = k^2\frac{d^2y}{dx^2}$

Substituting in ② we get.

$$\frac{t^2}{k^2}\cdot k^2\frac{d^2y}{dt^2} + \frac{t}{k}\, k\, \frac{dy}{dt} + (t^2 - n^2)y = 0$$

$\therefore \quad t^2\frac{d^2y}{dt^2} + t\frac{dy}{dt} + (t^2 - n^2)y = 0$

$$t^2\frac{d^2y}{dt^2} + t\frac{dy}{dt} + (t^2 - n^2)y = 0 \quad —③$$

Which is the same the same form as ① hence we have the following result.

$y = J_n(t)$ is the sol^n of eq^n ③

If $y = J_n(x)$ is the solut^n of eq^n ① (Bessel f^n) then $y = J_n(kx)$ is the solut^n of eq^n ②

$y = J_n(x)$ sol$\underline{n}$ of

$$x^2 \frac{d^2y}{dx^2} + x \frac{dy}{dx} + (x^2 - n^2) y = 0$$

$y = J_n(kx)$ sol$\underline{n}$ of

$$(kx)^2 \frac{d^2y}{d(kx)^2} + kx \frac{dy}{d(kx)} + ((kx)^2 - n^2) y = 0$$

3.1. Bessel equation for zero order

Bessel eq$\underline{n}$ of order (o) zero $n=0$

$$x^2 \frac{d^2y}{dx^2} + x \frac{dy}{dx} + (x^2 - 0) y = 0$$

dividing by x^2 we get

$$\boxed{\frac{d^2y}{dx^2} + \frac{1}{x} \frac{dy}{dx} + y = 0} \quad \text{—①} \qquad \boxed{\text{I}}$$

We assume as a solut$\underline{n}$ of this eq$\underline{n}$ an infinite series of this form

$$y = x^c (a_0 + a_1 x + a_2 x^2 + \cdots + a_n x^n), \ a_0 \neq 0$$

i.e

$$y = a_0 x^c + a_1 x^{c+1} + a_2 x^{c+2} + \cdots$$

$$\frac{dy}{dx} = a_0 c x^{c-1} + a_1 (c+1) x^c + a_2 (c+2) x^{c+1} + \cdots$$

$$\frac{d^2y}{dx^2} = a_0 c (c-1) x^{c-2} + a_1 (c+1) c x^{c-1} + a_2 (c+2)(c+1) x^c$$

$\left. \vphantom{\begin{array}{c}1\\2\\3\end{array}} \right] ②$

19

Substituting from y & its derivatives in eq^n ① we get.

$$a_0 C(C-1) x^{C-2} + a_1 (C+1) C x^{C-1} + a_2 (C+2)(C+1) x^{C} + \cdots$$
$$a_0 C \; x^{C-2} + a_1 (C+1) x^{C-1} + a_2 (C+2) \; x^{C} + \cdots$$
$$+ \; a_0 x^{C} + \qquad = 0$$
$$\qquad\qquad\qquad\qquad\qquad ③$$

$\underline{\underline{\text{Coeff}^t \text{ of } x^{C-2}}}:- \qquad a_0 (C^2 - C + C) = 0 \quad \therefore a_0 C^2 = 0$

$\qquad\qquad\qquad\qquad\qquad\qquad * \; \because a_0 \neq 0 \; \therefore C^2 = 0$

$\qquad\qquad\qquad\qquad\qquad\qquad \quad i.e \; \underline{C = 0, 0}$

This means that the indicial eq^n here has two equal zero roots. (index)

$\underline{\underline{\text{Coeff}^t \text{ of } x^{C-1}}} \qquad a_1 (C+1)^2 = 0 \qquad * \; \therefore \underline{a_1 = 0}, \; as C = 0$

$\underline{\underline{\text{coeff}^t \text{ of } x^{C}}} \qquad a_2 (C+2)^2 + a_0 = 0$

$\therefore$ In general we have $\boxed{a_{r+2} (C+r+2)^2 + a_r = 0}$

$$\therefore \quad a_{r+2} = \frac{-a_r}{(c+r+2)^2}$$

Put $\boxed{r=0}$ $\therefore a_2 = -\dfrac{a_0}{(c+2)^2}$ *

$\boxed{r=1}$ $\therefore a_3 = \dfrac{-a_1}{(c+3)^2} = 0$ * since $a_1 = 0$

And similarly all coeff^s with odd size suffix $= 0$

$\boxed{r=2}$

$$\therefore a_4 = -\frac{a_2}{(c+4)^2} = \frac{a_0}{(c+2)^2(c+4)^2} \quad *$$

$a_0 \neq 0$ $\qquad$ $\therefore a_1 = 0$

$a_2 = \dfrac{-a_0}{(c+2)^2}$ $\qquad$ $\therefore a_3 = 0$

$a_4 = \dfrac{a_0}{(c+2)^2(c+4)^2}$ $\qquad$ $\therefore a_5 = 0$

$a_6 = \dfrac{-a_0}{(c+2)^2(c+4)^2(c+6)^2}$ $\qquad$ $\therefore a_7 = 0$

Hence we get.

$$\boxed{y = a_0 x^c \left[1 - \frac{x^2}{(c+2)^2} + \frac{x^4}{(c+2)^2(c+4^2)} - \frac{x^6}{(c+2)^2(c+4)^2(c+6)^2} \right.}$$

———————— Ⅱ

First solution

Put $c = 0$ $\therefore y = a_0 \left[1 - \frac{x^2}{2^2} + \frac{x^4}{2^2 \cdot 4^2} - \frac{x^6}{2^2 \cdot 4^2 \cdot 6^2} + \cdots \right]$

21

The Second solut^n. The 2nd solut^n has various forms, one of these is due to (Neumann) & is obtained in the following way:

Substituting from ② in ① and taking $a_0 = 1$ we get.

(II) $\quad y_1 = a_0 x^c \left(1 + \frac{x^2}{2^2} + \frac{x^4}{2^2 4^2} \cdots \right]$

(I) $\quad \dfrac{d^2 y_1}{dx^2} + \dfrac{1}{x} \dfrac{dy_1}{dx} + y_1 = 0$

∴ i) $\dfrac{dy_1}{dx} = cx^{c-1} - \dfrac{(c+2)}{2^2} x^{c+1} + (c+4) \dfrac{x^{c+3}}{2^2 4^2}$.

.. ii) $\dfrac{d^2 y_1}{dx^2} = c(c-1)x^{c-2} - \dfrac{(c+2)(c+1)}{2^2} x^c + \dfrac{(c+4)(c+3)}{2^2 4^2} x^{c+2}$

∴ $\dfrac{d^2 y_1}{dx^2} + \dfrac{1}{x}\dfrac{dy_1}{dx} + y_1 = (c + c^2 - c)x^{c-2}$

$\qquad + \left[1 - \dfrac{c+2}{2^2} + \dfrac{(c+2)(c+1)}{2^2} \right] x^c$

$\qquad + \left[-\dfrac{1}{2^2} + \dfrac{(c+4)}{2^2 4^2} + \dfrac{(c+4)(c+3)}{2^2 4^2} \right] x^{c+2}$

∴ $\dfrac{d^2 y_1}{dx^2} + \dfrac{1}{x}\dfrac{dy_1}{dx} + y_1 = c^2 x^{c-2} + \left[1 - \dfrac{c+2}{2^2}[1 + c + 1] \right] x^c$

$\qquad + \left[-\dfrac{1}{2^2} + \dfrac{(c+4)}{2^2 4^2} [1 - (c+3)] \right] x^{c+2}$

substituting from ② in ① we get and taking $a_0 = 1$

⨿ $y_1 = a_0 x^c \left[1 - \dfrac{x^2}{(2+c)^2} + \dfrac{x^4}{(2+c)^2 (4+c)^2} \cdots \right]$

$\dfrac{dy_1}{dx} = c x^{c-1} - \dfrac{(c+2)}{(c+2)^2} x^{c+1} + \dfrac{(c+4)}{(c+2)(c+4)^2} x^{c+3} \cdots$

$\dfrac{d^2 y_1}{dx^2} = (c-1) c x^{c-2} - \dfrac{(c+2)(c+1)}{(c+2)^2} x^{c} + \dfrac{(c+4)(c+3)}{(c+2)^2(c+4)^2} x^{c+2}$

① $\dfrac{d^2 y}{dx^2} + \dfrac{1}{x} \dfrac{dy}{dx} + y =$

$= \left[c + c^2 - c \right] x^{c-2} + \left[1 - \dfrac{1}{c+2} - \dfrac{(c+1)}{(c+2)} \right] x^{c}$

$+ \left[-\dfrac{1}{(2+c)^2} + \dfrac{1}{(c+2)^2(c+4)} + \dfrac{c+3}{(c+2)(c+4)} \right] c^{+2}$

$= c^2 x^{c-2} + 0 + 0$

$\therefore \dfrac{d^2 y_1}{dx^2} + \dfrac{1}{x} \dfrac{dy_1}{dx} + y_1 = c^2 x^{c-2}$ ————Ⓐ

Differentiate the both sides w.r.t (c)

$\dfrac{\partial}{\partial c} \left(\dfrac{d^2 y_1}{dx^2} \right) + \dfrac{1}{x} \dfrac{\partial}{\partial c} \left(\dfrac{dy_1}{dx} \right) + \dfrac{\partial y_1}{\partial c} = c^2 x^{c-2} \log_e x + x^{c-2} \cdot 2c$

$\boxed{\dfrac{d^2}{dx^2} \left(\dfrac{\partial y_1}{\partial c} \right) + \dfrac{1}{x} \dfrac{d}{dx} \left(\dfrac{\partial y_1}{\partial c} \right) + \dfrac{\partial y_1}{\partial c} = c x^{c-2} \log x + 2 c x^{c-2}}$

at $c = 0$ Ⓛ

$\qquad \qquad = 0$

23

If we put $c=0$ the R.H.S is zero. is similar to the L.H.S of ① thus $\frac{\partial y_1}{\partial c}$ written instead (y_2).

∴ The 2nd solⁿ obtained by differentiating y_1 w.r.t (c) and putting $c=0$. The standard 2nd solⁿ due to (Weber) german is given by:—

$$Y_0(x) = \frac{2}{\pi}\left\{\log\left(\frac{x}{2}\right) + \gamma\right\} J_0(x) - \frac{2}{\pi}\sum_{\nu=1}^{\infty}\frac{(-1)^{\nu}\left(\frac{x}{2}\right)^{2\nu}}{(\nu!)^2}\left(1 + \frac{1}{2} + \frac{1}{3} + \cdots \frac{1}{\nu}\right)$$

Where $\gamma =$ Euler's constant $=$

$$= \lim_{n\to\infty}\left[\left(1 + \frac{1}{2} + \frac{1}{3} + \cdots + \frac{1}{n}\right) - \log n\right] = 0.5772..$$

The 2nd solⁿ is limited in use due to the existance of the term $\log\frac{x}{2}$ which becomes infinite ∞ when $x = 0$ hence the 2nd solⁿ can be used for regeons which do not contain the origin. general solⁿ :-

$$y = A J_0(x) + B Y_0(x)$$

$\downarrow$ $\downarrow$

1ˢᵗ solⁿ   2nd solⁿ

3.2. Properties of Bessel functions

<u>Properties of $J_0(x)$</u> :- $J_0(x) = 1 - \dfrac{x^2}{2^2} + \dfrac{x^4}{2^2.4^2} - \dfrac{x^6}{2^2.4^2.6^2} + \cdots$

i) The series of $J_0(x)$ is absolutely <u>convergent</u> for all values of x <u>real</u> or <u>complex</u>.

ii) The series of $J_0(x)$ contains only even 8^n power of x and hence $J_0(x)$ is an even f^n of x whose graph is <u>symmetrical</u> w.r.t the verticle axis

iii) $J_0(0) = 1$

iv) $J_0(x) \to 0$ as $x \to \infty$

v) The difference bet^n two consecutive zeroes of $J_0(x)$ tends to π as $x \xrightarrow{\text{tends}} \infty$

<u>NOTE</u> :-

By a zero of a function we mean a value of (x) which makes the function zero. in other words it is a root of the eq^n. we shall now prove the behaviour of $J_0(x)$ for large values of (x) write eq^n $\textcircled{I}$ in the form.

$$x \dfrac{d^2y}{dx^2} + \dfrac{dy}{dx} + xy = 0 \underline{\qquad} \textcircled{\scriptsize I}$$

and put $y = \dfrac{u}{\sqrt{x}}$

i.e ⅰ) $\quad y = U x^{-\frac{1}{2}}$

ⅱ) $\quad \dfrac{dy}{dx} = -\dfrac{1}{2} U x^{-\frac{3}{2}} + x^{\frac{1}{2}} \dfrac{dU}{dx}$

$$\dfrac{d^2 y}{dx^2} = \dfrac{3}{4} U x^{-\frac{5}{2}} - \dfrac{1}{2} x^{-\frac{3}{2}}\dfrac{dU}{dx} - \dfrac{1}{2} x^{-\frac{3}{2}}\dfrac{dU}{dx} + x^{-\frac{1}{2}}\dfrac{d^2 U}{dx}$$

ⅲ) $\quad \dfrac{d^2 y}{dx^2} = x^{-\frac{1}{2}}\dfrac{d^2 U}{dx} - x^{-\frac{3}{2}}\dfrac{dU}{dx} + \dfrac{3}{4} U x^{-\frac{5}{2}}.$

Substitute in ① we get

$$\dfrac{3}{4} U x^{-\frac{3}{2}} - x^{-\frac{1}{2}}\dfrac{dU}{dx} + x^{\frac{1}{2}}\dfrac{d^2 U}{dx^2}$$
$$-\dfrac{1}{2} U x^{-\frac{3}{2}} + x^{-\frac{1}{2}}\dfrac{dU}{dx} + U x^{\frac{1}{2}} = 0$$

$$= \dfrac{1}{4} U x^{-\frac{3}{2}} + x^{\frac{1}{2}}\dfrac{d^2 U}{dx^2} + U x^{\frac{1}{2}} = 0 \quad —②$$

multiply by $x^{-\frac{1}{2}}$

$$\therefore \quad \dfrac{1}{4} U x^{-2} + \dfrac{d^2 U}{dx^2} + U = 0$$

$\therefore$ i.e. $\quad \dfrac{d^2 U}{dx^2} + \left(1 + \dfrac{1}{4x^2}\right) U = 0 \quad —③$

Since we are concerned $\cancel{\text{it}}$ with a large value of (x)

$\therefore$

$$\dfrac{d^2 U}{dx^2} + U = 0$$

i.e sol^n $\quad U = C \cos(x - \lambda)$ $\qquad$ instead of $A \sin + B \cos$

$\therefore$ $y = \dfrac{C \cos(x - \lambda)}{\sqrt{x}}$

Since the cosine f^n is bounded hence $y = 0$ as $x \to \infty$
i.e

$$\boxed{J_0(x) \to 0 \text{ as } x \to \infty}$$

$y = 0$ when $\cos(x - \lambda) = 0$ showing that the difference
bet^n two consecutive $\dot{\imath} \cdot e$ zeros tends to π as x tends to
∞. In fact the curve $J(x)$ ressembles $\dot{\imath}\circ$ at
Adampped cosine $\circ_3$ curve.
as when $\sqrt{x} \to \infty$ $y = 0$

$$\therefore y = \cos x$$

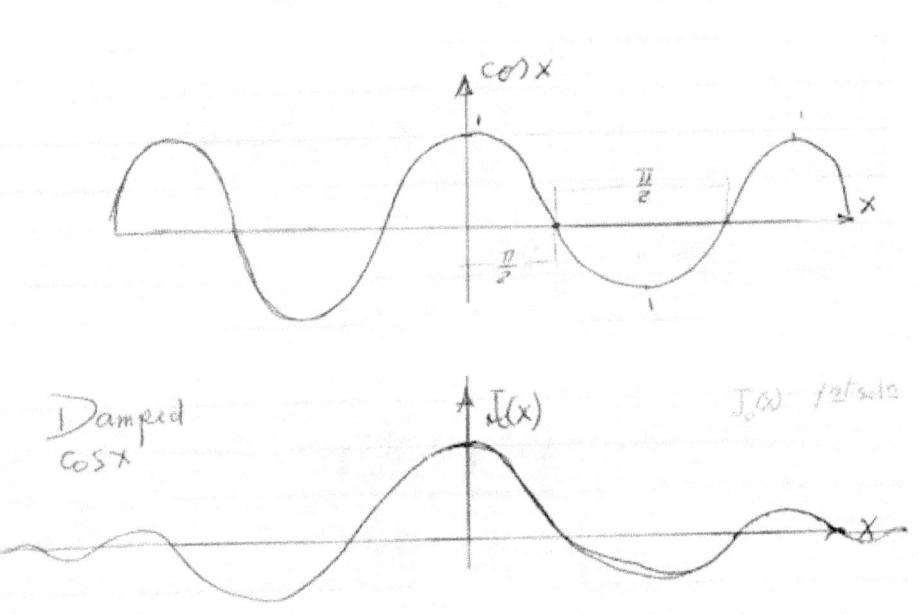

Damped
cos x

$J_0(x)$ 1st soln

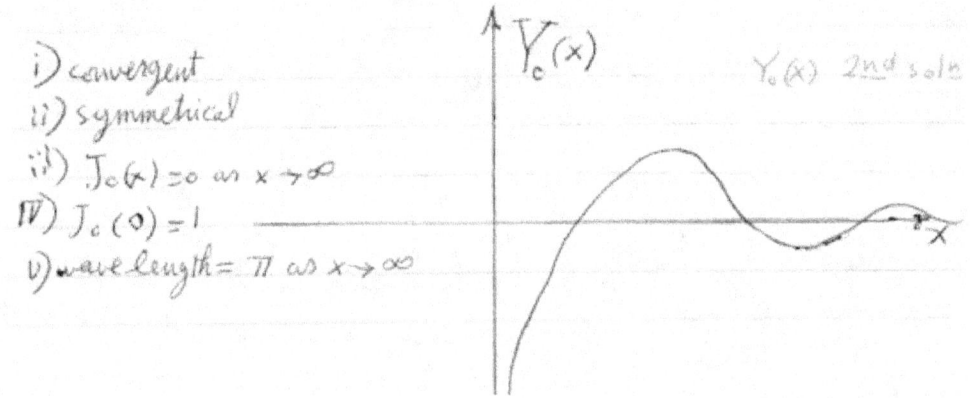

i) convergent
ii) symmetrical
iii) $J_0(x) = 0$ as $x \to \infty$
iV) $J_0(0) = 1$
v) wave length $= \pi$ as $x \to \infty$

$d \setminus 3 \setminus ^{19}72$

Zeros of $J_0(x)$. $x = 2.405; 5.52; 8.654; 11.792$

3.3. Bessel functions of order one (n = 1)

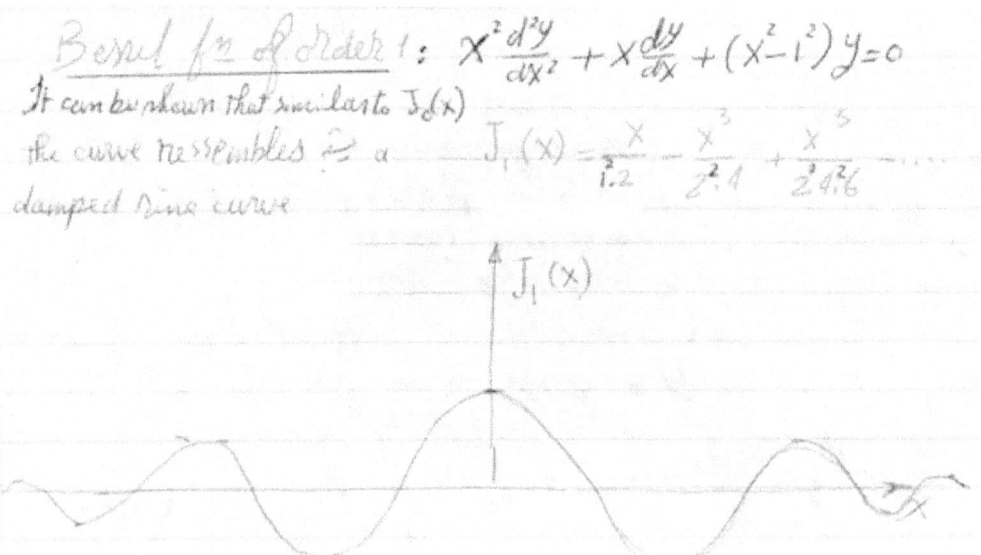

Bessel fn of order 1: $x^2 \dfrac{d^2 y}{dx^2} + x \dfrac{dy}{dx} + (x^2 - 1^2) y = 0$

It can be shown that similar to $J_0(x)$ the curve ressembles a damped sine curve

$J_1(x) = \dfrac{x}{1.2} - \dfrac{x^3}{2^2 \cdot 4} + \dfrac{x^5}{2^2 4 \cdot 6} - \cdots$

3.4. Relationships between Bessel functions of orders zero and one

Two differential formulae connecting Bessel functions follow.

Two fundamental formulæ connecting $J_0(x)$ & $J_1(x)$

$$J_0(x) = 1 - \frac{x^2}{2^2} + \frac{x^4}{2^2 \cdot 4^2} - \frac{x^6}{2^2 \cdot 4^2 \cdot 6^2}$$

$$J_1(x) = \frac{x}{1 \cdot 2} - \frac{x^3}{2^2 \cdot 4} + \frac{x^5}{2^2 \cdot 4^2 \cdot 6} - \frac{x^7}{2^2 \cdot 4^2 \cdot 6^2 \cdot 8}$$

$$\frac{d}{dx}\{J_0(x)\} = -\frac{x}{2} + \frac{x^3}{2^2 \cdot 4} - \frac{x^5}{2^2 \cdot 4^2 \cdot 6} = -J_1(x) \quad\text{——①}$$

from this relat^n it follows that zeros of $J_1(x)$ that is the value of x which make $J_1(x) = 0$ are at the same time point of maxima & minima of $J_0(x)$. the fig. shows that zeros of $J_0(x)$ & $J_1(x)$ separates one another, that is there are no two zeros of $J_0(x)$ which lie bet^n 2 zeros of $J_1(x)$. The second fundamental relat^n bet^n $J_0(x)$ & $J_1(x)$ is

$$\frac{d}{dx}\{x J_1(x)\} = x J_0(x) \quad\longrightarrow ②$$

0

example : Prove that $\int_0^1 x J_0(\alpha x) dx = \frac{1}{\alpha} J_1(\alpha)$

sol^n :

$\because \int x J_0(x) dx = x J_1(x)$ change x into αx

$\therefore \int \alpha x J_0(\alpha x) d(\alpha x) = \alpha x J_1(\alpha x)$

divide by α^2 $\therefore \int_0^1 x J_0(\alpha x) dx = \frac{1}{\alpha}[x J_1(\alpha x)]_0^1 = \frac{1}{\alpha} J_1(\alpha) - 0$

example 2 . if $U_n = \int x^n J_0(x) dx$, prove the reduct^n formula

$$U_n = x^n J_1(x) + (n-1) x^{n-1} J_0(x) - (n-1)^2 U_{n-2}$$

sol^n key to sol^n

$$U_n = \int x^{n-1} \cdot x J_0(x) dx = \int x^{n-1} d\{x J_1(x)\}$$

Introduction to integration by partial.

$$U_n = x^n \cdot J_1(x) - \int x J_1(x)\, dx^{n-1}$$

$$= x^n \cdot J_1(x) - (n-1)\int x^{n-1} J_1(x)\, dx$$

$$= x^n \cdot J_1(x) + (n-1)\int x^{n-1}\, d\{J_0(x)\}$$

$$\underbrace{\qquad}_{\text{another time}}$$

$$x^{n-1} = z$$
$$(n-1)x^{n-2}\, dx = dz$$

$$= x^n \cdot J_1(x) + (n-1) x^{n-1} \cdot J_0(x) - (n-1)^2 \int x^{n-2} \cdot J_0 x\, dx$$

$$\therefore U_n = x^n J_1(x) + (n-1) x^{n-1} J_0(x) - (n-1)^2 U_{n-2}$$

3.5. Lammel's integrals

Lommel's integrals

We already known that one solution of $x \dfrac{d^2 y}{dx^2} + \dfrac{dy}{dx} + k^2 x y = 0$
(Bessel of zero order) is $y = J_0(kx)$

hence one soln of $\quad x u'' + u' + \alpha^2 x u = 0$. is $\quad u = J_0(\alpha x)$ —①

similarly one soln of $\quad x v'' + v' + \beta^2 x v = 0$ is $v = J_0(\beta x)$ —②

multiply eqn ① by v & eqn ② by u & substract.

$$x\{u'' v - v'' u\} + \{u' v - v' u\} + \{\alpha^2 - \beta^2\} x u v = 0$$

i.e:
$$\frac{d}{dx}\{x(u' v - v' u)\} = (\beta^2 - \alpha^2) x u v$$

as

$$\underline{u' v} + \underline{x u'' v} + x u' v' - \underline{v' u} - \underline{x v'' u} - x v' u' = (u' v - v' u) + x(u'' v - v'' u)$$

$$= -(\alpha^2 - \beta^2) x u v$$

$$\therefore \int x(u v)\, dx = \frac{x(u' v - v' u)}{\beta^2 - \alpha^2}, \qquad u v \text{ عند}$$

from ① & ②

$$\therefore \int x \cdot J_0(\alpha x) J_0(\beta x)\, dx = \frac{x\{\alpha J_0'(\alpha x) J_0(\beta x) - \beta J_0(\beta x) J_0'(\alpha x)\}}{(\beta^2 - \alpha^2)}$$

Dashes means that differentiation of J_0 w.r.t (αx) & (βx)

observe $u' = \dfrac{d}{dx} J_0(\alpha x) = \alpha \dfrac{d}{d\alpha x} J_0(\alpha x) = \alpha J_0'(\alpha x)$

& so $v' = $ _____ $= \beta J_0'(\beta x)$

$\therefore$

$$\int_0^1 x J_0(\alpha x) \cdot J_0(\beta x)\, dx = \dfrac{\alpha J_0'(\alpha) J_0(\beta) - \beta J_0'(\beta) \cdot J_0(\alpha)}{\beta^2 - \alpha^2} \quad \boxed{3}$$

note $J_0(0) = 1 \qquad 1 - 1 = 0$

We now consider the particular but important case when α, β are zeros of $J_0(x)$ that is $J_0(\alpha) = 0$, $J_0(\beta) = 0$

case I $\quad \alpha \neq \beta$, $\qquad$ orthogonal relation with

$$\int_0^1 x J_0(\alpha x) \cdot J_0(\beta x)\, dx = 0 \qquad \cancel{\text{fix}}$$

i.e

$\sqrt{x}\, J_0(\alpha x)$ & $\sqrt{x}\, J_0(\beta x)$ are orthogonal fns

in interval $(0,1)$

case II $\quad \alpha = \beta$ we use the method of indetermined form.

Here the R.H.S of $\boxed{3}$ is of the indetermined form $\dfrac{0}{0}$
we therefore fix (α) & let (β) tends to $\longrightarrow \alpha$ we
differentiate both numerator & denominator &
take the limit when $\beta \longrightarrow \alpha$ $\quad$ w.r.t.(β)

$\lim\limits_{\beta \to \alpha} R.H.S = \lim\limits_{\beta \to \alpha} \dfrac{\alpha J_0'(\alpha) J_0'(\beta)}{2\beta}$ $\qquad \alpha = \text{const}.$

$$= \frac{1}{2} J_0'^2(\alpha) = \frac{1}{2} J_1^2(\alpha)$$

since $J_0'(\alpha) = -J_1(\alpha)$

$$\therefore \int_0^1 x\, J_0^2(\alpha x)\, dx = \frac{1}{2} J_1^2(\alpha)$$

$$\lim_{\beta \to \alpha} \frac{\alpha J_0'(\alpha) J_0'(\beta) - \beta J_0'(\beta) J_0(\alpha)}{2\beta}$$

Hence α, β are roots of eye $\therefore J_0'(\alpha) = 0$

$J_0(\alpha) = 0$

$J_0(\alpha) \neq 0$

Hence we have the following fundamental result:-
if α & β are zeros of $J_0(x)$ i.e $J_0(\alpha) = 0$ & $J_0(\beta) = 0$

3.6. Fourier-Bessel expansions of zero order

Then

$$\int_0^1 x\, J_0(\alpha x) J_0(\beta x)\, dx = 0 \qquad \alpha \neq \beta$$

$$\int_0^1 x\, J_0^2(\alpha x)\, dx = \frac{1}{2} J_1^2(\alpha) \qquad \alpha = \beta$$

We now make use of this result in obtaining the <u>Fourier-Bessel</u> expansion of a given fn $F(x)$ in the range of $(0,1)$.

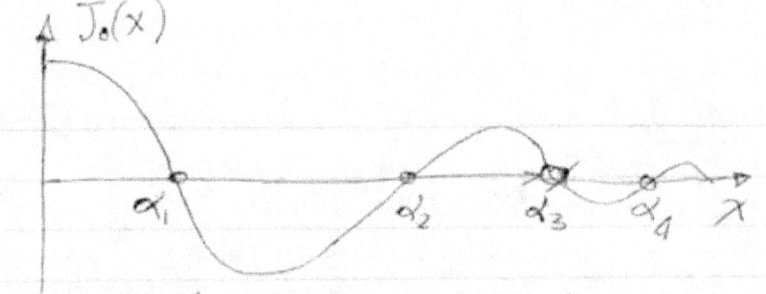

Let $\alpha_1, \alpha_2, \alpha_3 \ldots$ be the positive zeros of $J_0(x)$ arranged in ascending order vrew of magnitude and suppose that $F(x)$ can be expanded in the form.

$$f(x) = A_1 J_0(\alpha_1 x) + A_2 J_0(\alpha_2 x) + \cdots A_n J_0(\alpha_n x)$$

in the interval $(0,1)$ to obtain __A_n__. multiply both sides by $x J_0(\alpha_n x)$ & integrate bet^n 0 & 1.

$$\int_0^1 x J_0(\alpha_n x) \, f(x) \, dx = A_n \int_0^1 x J_0^2(\alpha_n x) \, dx$$

$$= A_n \cdot \frac{1}{2} J_1^2(\alpha_n)$$

All other integrals vanish according to the property of orthogonality, $\overset{u}{\sim}$ i.e. $\sqrt{x} \, J_0(\alpha_1 x) \cdot \sqrt{x} \, J_0(\alpha_2 x) = 0$

$$\therefore A_n = \frac{2}{J_1^2(\alpha_n)} \int_0^1 x \, J_0(\alpha_n x) \, f(x) \, dx.$$

ex. Expand unity in a series of bessel f^n of order zero.
i.e $1 = A_1 J_0(\alpha_1 x) + A_2 J_0(\alpha_2 x) + \cdots + A_n J_0(\alpha_n x)$

$$A_n = \frac{2}{J_1^2(\alpha_n)} \int_0^1 x \, J_0(\alpha_n x) \, dx$$

$$= \frac{2}{J_1^2(\alpha_n)} \cdot \frac{1}{\alpha_n} J_1(\alpha_n) = \frac{2}{\alpha_n} \frac{1}{J_1(\alpha_n)}$$

$$\therefore 1 = \frac{2}{\alpha_1} \frac{J_0(\alpha_1 x)}{J_1(\alpha_1)} + \frac{2}{\alpha_2} \frac{J_0(\alpha_2 x)}{J_1(\alpha_2)} + \cdots + \frac{2}{\alpha_n} \frac{J_0(\alpha_n x)}{J_1(\alpha_n)} + \cdots$$

$$i.e \quad 1 = \sum_{n=1}^{\infty} \frac{2}{\alpha_n} \frac{J_0(\alpha_n x)}{J_1(\alpha_n)}$$

3.7. Vibration of uniformly stretched membrane

11 /3/ 1972

APPLICATION

Free vibrations of uniformly stretched membranes:-

1. Constant tension
2. Regular membrane

we consider a perfectly flexible membrane. that is a membrane such that the stress accross any line drawn on the membrane is a tension normal to the line at every point. and lies in the tangent plane to the membrane at that point (there is no shear). we shall now assume

that the membrane is plane in its equilibrium position under the action of a uniform tension T per unit length over the boundry.

Consider the equilibrium of the element ABCD & resolve // parallel to x-axis.

$$T'dy = Tds. \cos \psi.$$
$$= T. ds \cos \psi = dy$$
$$\therefore T' = T \quad \text{therefore it is uniformly stretched membrane.}$$

34

i.e. the tension per unit length accross any line drawn on the membrane $T dy$ is constant & equal to its value along the boundry.

we now obtain the differential eq^{n} of the motion. we make the following assumptions:-

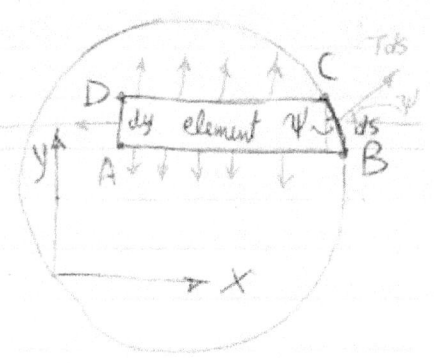

① we shall assume that the membrane is horizontal. the lateral displacement is in the vertical direct^{n} which will take as the z direct^{n}

② we shall neglect the effect of gravity.

③ we shall assume small lateral displacement (z) as well as small gradients.

. the tension accross $\boxed{EF}$ is $T dy$

. its component in the oz-direction is $T dy \cos \alpha = T dy . \dfrac{dz}{AB}$ since it is small angle.

$\therefore T dy \cos \alpha = T dy \dfrac{\partial z}{\partial x}$ ——①

the component of tension across $\boxed{BC}$ in the oz-direct^{n} is

② $T dy \dfrac{\partial z}{\partial x} + \dfrac{\partial}{\partial x}\left(T dy \dfrac{\partial z}{\partial x}\right) \dfrac{\partial x}{2}$

similarly the component of the tension $\boxed{AD}$ in the direction zo

③ $T dy \dfrac{\partial z}{\partial x} - \dfrac{\partial}{\partial x}\left(T dy \dfrac{\partial z}{\partial x}\right) \dfrac{dx}{2}$

the resulted force in the direct^{n} oz is ②-③

$$\frac{\partial}{\partial x}\left(T dy \frac{\partial z}{\partial x}\right) dx = T \frac{\partial^2 z}{\partial x^2} dy\, dx \quad\text{———}\quad (I)$$

we have a similar contribution from the forces across AB & CD & hence the total force acting on the membrane in the direction OZ

$$T\left(\frac{\partial^2 z}{\partial x^2} + \frac{\partial^2 z}{\partial y^2}\right) dy\, dx \quad\text{———}\quad (II)\quad\text{This should}$$

be equal to

$$\sigma\quad dx\, dy \frac{\partial^2 z}{\partial t^2} \quad\text{———}\quad (III)$$

where σ is the surface density i.e the mass per unit area & $\frac{\partial^2 z}{\partial t^2}$ = acceleration.

$$\therefore\quad \sigma \frac{\partial^2 z}{\partial t^2} = T\left(\frac{\partial^2 z}{\partial x^2} + \frac{\partial^2 z}{\partial y^2}\right) \qquad\text{but } c = \sqrt{\frac{T}{\sigma}}$$

$$\therefore\quad \frac{\partial^2 z}{\partial t^2} = c^2\left[\frac{\partial^2 z}{\partial x^2} + \frac{\partial^2 z}{\partial y^2}\right]$$

we now consider a circular membrane of radius (a) & let us take the centre of the circle as the origin. Transforming the partial differential eqns to be polar coordinate we get.

put $dx = dr$
$dy = r\, d\theta$.

$$\therefore\quad \frac{\partial}{\partial x}\left(T dy \frac{\partial z}{\partial x}\right) dx = \frac{\partial}{\partial r}\left(T r d\theta \frac{\partial z}{\partial r}\right) dr$$
$$= T\left(r \frac{\partial^2 z}{\partial r^2} + \frac{\partial z}{\partial r}\right) dr\, d\theta$$

$$\therefore\quad \frac{\partial}{\partial y}\left[T dx \frac{\partial z}{\partial y}\right] dy = \frac{\partial}{r\partial\theta}\left[T dr \frac{\partial z}{r\partial\theta}\right] r d\theta$$

$$= \frac{T}{r}\left(\frac{\partial^2 z}{\partial \theta^2}\right) dr\, d\theta$$

$$\sigma \frac{\partial^2 z}{\partial t^2} \overset{dxdy}{=} T\left(\frac{\partial^2 z}{\partial x^2} + \frac{\partial^2 z}{\partial y^2}\right)^{dxdy} = T\left(r\frac{\partial^2 z}{\partial r^2} + \frac{\partial z}{\partial r} + \frac{1}{r}\frac{\partial^2 z}{\partial \theta^2}\right) dr\, d\theta$$

$$= T\left(\frac{\partial^2 z}{\partial r^2} + \frac{1}{r}\frac{\partial z}{\partial r} + \frac{1}{r^2}\frac{\partial z}{\partial \theta^2}\right) r\, d\theta\, dr$$

$$\therefore \sigma \frac{\partial^2 z}{\partial t^2} = T\left(\frac{\partial^2 z}{\partial r^2} + \frac{1}{r}\frac{\partial z}{\partial r} + \frac{1}{r^2}\frac{\partial^2 z}{\partial \theta^2}\right)$$

$$\boxed{\frac{\partial^2 z}{\partial t^2} = c^2\left(\frac{\partial^2 z}{\partial r^2} + \frac{1}{r}\frac{\partial z}{\partial r} + \frac{1}{r^2}\frac{\partial z}{\partial \theta^2}\right)}$$

Assuming symmetry about the polar axis of the membrane i.e the axis through the centre normal to the membrane. The Z is independent of θ & eq^n of motion reduces to:

shall $\quad \dfrac{\partial^2 z}{\partial t^2} = c\left(\dfrac{\partial^2 z}{\partial r^2} + \dfrac{1}{r}\dfrac{\partial z}{\partial r}\right)$

we now assume a normal mode of vibration.

$$Z = R\cos(\omega t - \epsilon) \qquad \text{where } R = f(r)$$

In which all particles execute is Simple Harmonic Oscilations of the same frequency & passing simultaneously through their respective positions of equilibrium. Substituting in the differential eq^n & dividing by the time factor we get.

$$-\omega^2 R = c^2\left(\frac{d^2 R}{dr^2} + \frac{1}{r}\frac{dR}{dr}\right) = c^2\left(\frac{d^2 R}{dr^2} + \frac{1}{r}\frac{dR}{dr}\right)$$

i.e. $\dfrac{d^2R}{dr^2} + \dfrac{1}{r}\dfrac{dR}{dr} + \dfrac{\omega^2}{c^2}R = 0$

This is a Bessel fn. d. eq$\underline{n}$ of zero order given as sol$\underline{n}$

$$R = A J_0\left(\dfrac{\omega}{c}r\right) + B Y_0\left(\dfrac{\omega}{c}r\right)$$

$$\therefore Z = \left\{A J_0\left(\dfrac{\omega}{c}r\right) + B Y_0\left(\dfrac{\omega}{c}r\right)\right\}\cos(\omega t - \varepsilon)$$

Since Y_0 becomes ∞ (infinite) at the origin while z is finite then $B = 0$

$$\therefore \quad Z = A J_0\left(\dfrac{\omega r}{c}\right)\cos(\omega t - \varepsilon)$$

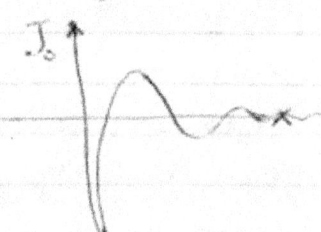

Initial conditions,

Since the boundry is fixed then $Z = 0$ when $r = a$ for all times

$\therefore \quad Z = 0 = J_0\left(\dfrac{\omega a}{c}\right)$

This means that $\dfrac{\omega a}{c}$ is a zero of $J_0(x)$

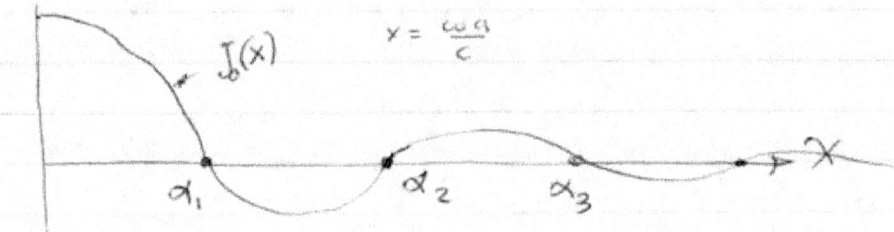

$$\frac{\omega_1 a}{c} = \alpha_1 \quad , \quad \frac{\omega_2 a}{c} = \alpha_2 \quad , \quad \frac{\omega_2 a}{c} = \alpha_3$$

$$Z_1 = A_1 J_0\left(\alpha_1 \frac{r}{a}\right) \cos\left(\frac{c\alpha_1}{a} t - \epsilon_1\right)$$

this gives the 1$\underline{st}$ normal mode of vibrations. The 2$\underline{nd}$ mode is given by,

$$Z_2 = A_2 J_0\left(\alpha_2 \frac{r}{a}\right) \cos\left(\frac{c\alpha_2}{a} t - \epsilon_2\right)$$

The General motion is a linear combination of the normal modes of Vibrations.

i.e.

$$z = \sum_{n=1}^{\infty} A_n J_0\left(\alpha_n \frac{r}{a}\right) \cos\left(\frac{c\alpha_n}{a} t - \epsilon_n\right)$$

the arbitrary constant A_n & ϵ_n are determined from initial conditions. i.e the initial shape of the membrane & the initial velocity & here we can use the fourier Bessel expansion of zero order.

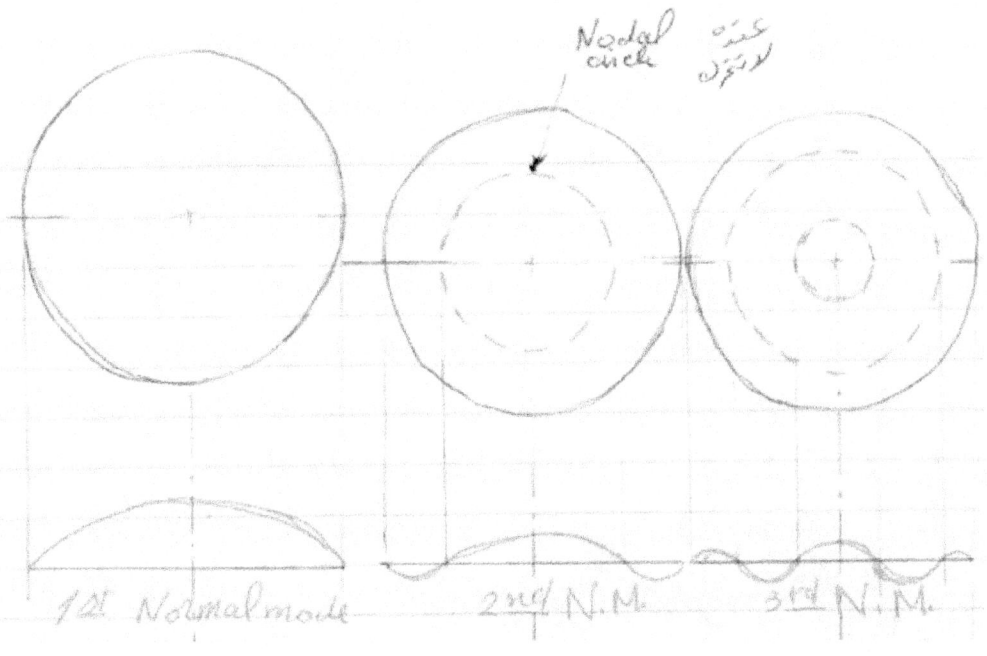

Nodal circle الدائرة العقدية

1$\underline{st}$ Normal mode   2$\underline{nd}$ N.M   3$\underline{rd}$ N.M.

39

the dynamic deformation surface of the membrane at any instant is a surface of revolution generated by rotating a J_0 curve about the polar Axis of the membrane.

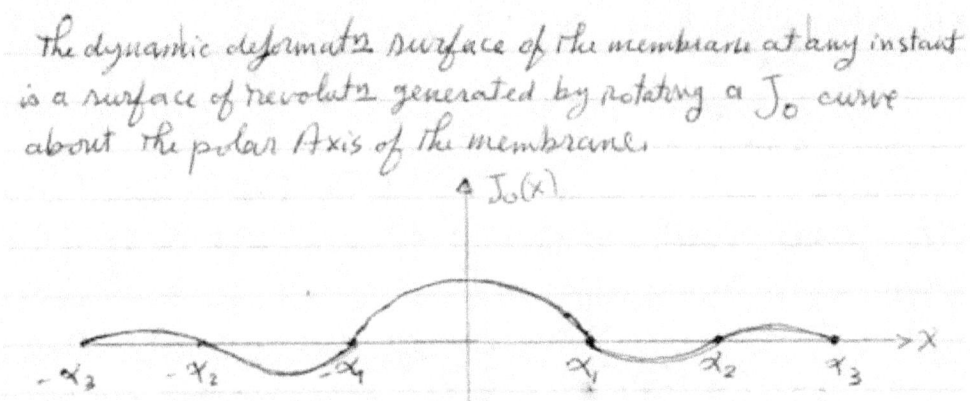

3.8. <u>Application</u> of Bessel functions on conduction of heat

$18 \backslash 3 \backslash ^{1972}$

APPLICATION of Bessel fns of zero order to problems of conduction of Heat

$$\rho, C, K, U$$

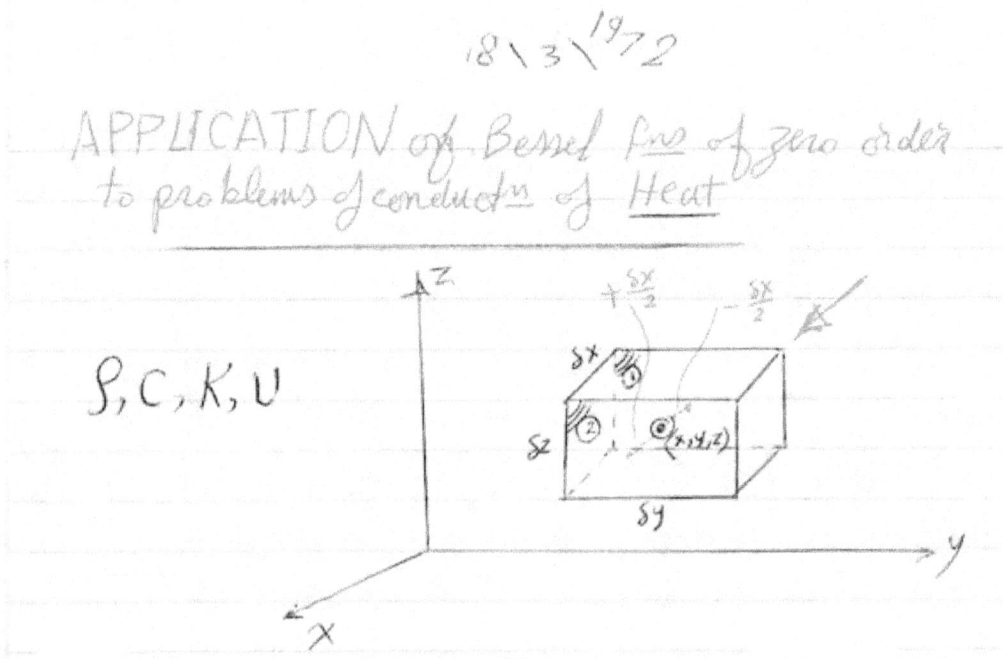

40

Let ρ be the density of the conducting medium. C is the specific heat. & K be its thermal conductivity.

Consider an element of solid in the form of a rectangular parallelepiped with centre (x, y, z) & edges $\delta x, \delta y, \delta z$ & Let U be a temperature at the centre (x, y, z) then U is a fn of x, y, z & t.

Quantity of heat flowing into the element through the face ① in time δt is

$$- K \, \delta y \, \delta z \left[\frac{\partial U}{\partial x} - \frac{1}{2} \frac{\partial^2 U}{\partial x^2} \, \delta x \right] \delta t \longrightarrow$$

Similarly the quantity of heat flowing out of the element from face ② in δt

$$- K \, \delta y \, \delta z \left[\frac{\partial U}{\partial x} + \frac{1}{2} \frac{\partial^2 U}{\partial x^2} \, \delta x \right] \delta t \longrightarrow$$

Gain of heat in the element.

$$= \text{input} - \text{output} = K \, \delta x \, \delta y \, \delta z \, \delta t \, \frac{\partial^2 U}{\partial x^2} \longrightarrow$$

Adding two similar expressions for the total gain of heat

$$= K \, \delta x \, \delta y \, \delta z \, \delta t \left[\frac{\partial^2 U}{\partial x^2} + \frac{\partial^2 U}{\partial y^2} + \frac{\partial^2 U}{\partial z^2} \right]$$

this should be equal to

$$= \rho \, \delta x \, \delta y \, \delta z \cdot c \frac{\partial U}{\partial t} \cdot \delta t.$$

$$\therefore \quad \frac{\partial^2 U}{\partial x^2} + \frac{\partial^2 U}{\partial y^2} + \frac{\partial^2 U}{\partial z^2} = \frac{\rho c}{K} \frac{\partial U}{\partial t}$$

Put $\frac{\rho c}{K} = \frac{1}{k}$ where k is the thermal Diffusivity

41

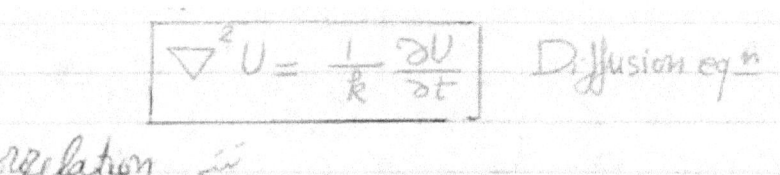

$$\boxed{\nabla^2 U = \frac{1}{k} \frac{\partial U}{\partial t}} \qquad \text{Diffusion eq}^n$$

correlation :-

if the flow of heat is steady then U is independent of the time & the heat eq^n reduces to (that uses to)

$$\nabla^2 U = 0 \qquad \text{steady flow}$$

i.e. the temperature satisfies Laplace's eq^n.

example

A long solid circular cylinder, has radius 20 centimetre one base of the cylinder is kept at fixed temp $100°c$ while the lateral surface is & the other base are kept at zero $0°c$. when the flow of heat is steady find the temp distribut^n in side the cylinder.

(A suitable coordinates to the form given.)

So/^{n}

Since the flow of heat is steady. U satisfies Laplace's eq^n

$$\nabla^2 U = \frac{\partial^2 U}{\partial x^2} + \frac{\partial^2 U}{\partial y^2} + \frac{\partial^2 U}{\partial z^2} = 0$$

$$= \frac{\partial^2 U}{\partial r^2} + \frac{1}{r}\frac{\partial U}{\partial r} + \frac{1}{r^2}\frac{\partial^2 U}{\partial \theta^2} + \frac{\partial^2 U}{\partial z^2}$$

$$= 0$$

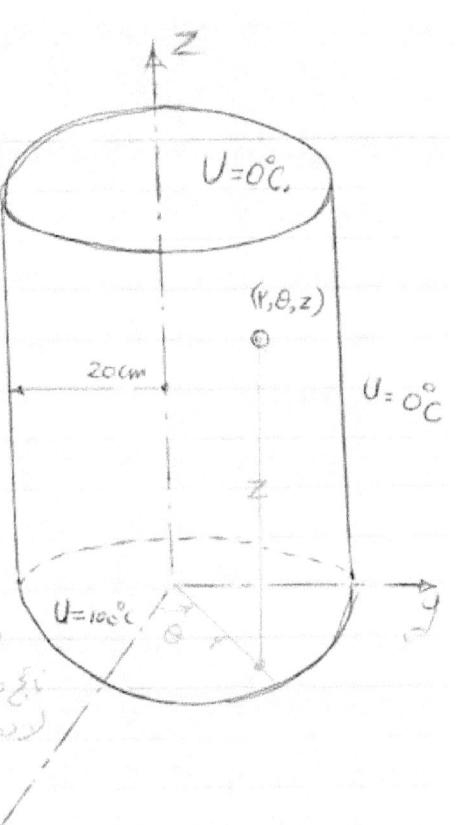

Since we have symmetry about z-axis, ∴
U is independent of θ & we get.

$$\frac{\partial^2 U}{\partial r^2} + \frac{1}{r}\frac{\partial U}{\partial r} + \frac{\partial^2 U}{\partial z^2} = 0$$

we assume the soln of this eqn.

$\quad U = f(r, z)$

$\quad U = RZ$

when R is a fn of
(r) only & Z is a fn of (z) only

$$Z\frac{d^2 R}{dr^2} + Z\frac{1}{r}\frac{dR}{dr} + R\frac{d^2 Z}{dz^2} = 0 \qquad \text{dividing by } RZ$$

$$\frac{1}{R}\left(\frac{d^2 R}{dr^2} + \frac{1}{r}\frac{dR}{dr}\right) + \frac{1}{Z}\frac{d^2 Z}{dz^2} = 0$$

$$\therefore \quad \frac{1}{R}\left(\frac{d^2 R}{dr^2} + \frac{1}{r}\frac{dR}{dr}\right) = -\frac{1}{Z}\frac{d^2 Z}{dz^2} = \text{constant say} = -\lambda^2$$

fn of r only = fn of z only.   Now that r & z which are
independent variables are equal it must be that equal constant
say $(-\lambda^2)$.

$$\therefore \quad \frac{1}{R}\left(\frac{d^2R}{dr^2} + \frac{1}{r}\frac{dR}{dr}\right) = -\lambda^2 \quad \& \quad \frac{1}{Z}\frac{d^2Z}{dz^2} = \lambda^2$$

$\therefore$ the differential partial eqn is divide to 2-exact eqns

$$\therefore \quad \frac{d^2R}{dr^2} + \frac{1}{r}\frac{dR}{dr} + \lambda^2 R = 0 \qquad \text{Bessel eqn} \longrightarrow \text{①}$$
$$\text{of zero order}$$

$$\frac{d^2Z}{dz^2} - \lambda^2 Z = 0 \longrightarrow \text{②}$$

Eqn ① is a Bessel diff eqn of zero order.

$$R = A J_0(\lambda r) + B Y_0(\lambda r) \longrightarrow \text{③}$$
$$Z = C e^{\lambda z} + D e^{-\lambda z} \longrightarrow \text{④}$$

$$\boxed{U = RZ = \left[A J_0(\lambda r) + B Y_0(\lambda r)\right]\left[C e^{\lambda z} + D e^{-\lambda z}\right]}$$

The term Y_0 is inadmissible of $\frac{z-3}{3}$ since it becomes infinite at $r=0$ while the temp is finite for pts on the axis of the cylinder $\therefore B = 0 \to$ ①

since the cylinder is long we may assume that $U \to 0$ as $z \to \infty$

$\therefore \qquad C = 0 \to$ ② (which is useful term is remained the useless terms are eliminated)'

$\therefore \quad C e^{\lambda z} = 0$

$\therefore$ the soln

$$U = A J_0(\lambda r) e^{-\lambda z} \qquad D, A \text{ another constant } A$$

① when $r = 20$ $U = 0$ for all z $\therefore 0 = J_0(20\lambda) = 0$

$\therefore 20\lambda = d_1, d_2, d_3 \dots$ when the d's are the $+ve$ zeros of $J_0(x)$ $\therefore \lambda_n = \dfrac{\alpha_n}{20}$

$$\therefore U = \sum_{n=1}^{n=\infty} A_n J_0\left(\alpha_n \frac{r}{20}\right) e^{-\frac{\alpha_n z}{20}}$$

(2) when $Z = 0$ $\quad U = 100°C$

$\therefore 100 = \sum_{n=1}^{\infty} A_n J_0 \left(\alpha_n \frac{r}{20} \right)$ $\qquad$ put $\frac{r}{20} \Rightarrow x$

Since r varies from 0 to 20, x varies from 0 to 1

$\therefore 100 = \sum_{n=1}^{\infty} A_n J_0 (\alpha_n x)$

$A_n = \dfrac{2}{J_1^2(\alpha_n)} \displaystyle\int_0^1 100 \, x \, J_0(\alpha_n x) \, dx$.

$= \dfrac{200}{J_1^2(\alpha_n)} \cdot \dfrac{1}{\alpha_n} J_1(\alpha_n)$

$A_n = \dfrac{200}{\alpha_n} \cdot \dfrac{1}{J_1(\alpha_n)}$

$\therefore U = \sum_{n=1}^{\infty} \dfrac{200}{\alpha_n} \cdot \dfrac{1}{J_1(\alpha_n)} \cdot J_0\left(\alpha_n \frac{r}{20}\right) e^{-\frac{\alpha_n Z}{20}}$

$= U(r, Z)$ temperature at pt (r, Z)

$U = \sum_{n=1}^{\infty} \dfrac{200}{\alpha_n} \cdot \dfrac{1}{J_1(\alpha_n)} \cdot J_0\left(\alpha_n \frac{r}{20}\right) e^{-\frac{\alpha_n Z}{20}}$

Note: Bessel functions are sometimes called cylinder functions or cylindrical harmonics.

3.9. Modified Bessel function of zero order

The Modified Bessel fns of zero Order

this have solns of the modified Bessel d. eqn of zero order

$$x \frac{d^2 y}{dx^2} + \frac{dy}{dx} - xy = 0 \quad \text{——①}$$

compared this with

$$x \frac{d^2 y}{dx^2} + \frac{dy}{dx} + k^2 xy = 0 \quad \text{——②} \qquad k^2 = -1 \quad k = i$$

Hence solns of ① are

$$J_0(ix) \quad \& \quad Y_0(ix)$$

$$J_0(x) = 1 - \frac{x^2}{2^2} + \frac{x^4}{2^2 \cdot 4^2} - \frac{x^6}{2^2 \cdot 4^2 \cdot 6^2} + \cdots$$

Denoting $J_0(ix)$ by $I_0(x)$ where $I_0(x)$ is known as the modified Bessel fns of zero order of the 1st kind we get.

$$I_0(x) = J_0(ix) = 1 - \frac{(ix)^2}{2^2} + \frac{(ix)^4}{2^2 \cdot 4^2} - \frac{(ix)^6}{2^2 \cdot 4^2 \cdot 6^2} + \frac{(ix)^8}{2^2 \cdot 4^2 \cdot 6^2 \cdot 8^2}$$

1st soln

$$\boxed{J_0(ix) = I_0(x) = 1 + \frac{x^2}{2^2} + \frac{x^4}{2^2 \cdot 4^2} + \frac{x^6}{2^2 \cdot 4^2 \cdot 6^2} + \frac{x^8}{2^2 \cdot 4^2 \cdot 6^2 \cdot 8^2}}$$

The standard 2nd soln Denoted by $K_0(x)$ is given by.

$$K_0(x) = \frac{\pi i}{2} \left[J_0(ix) + i Y_0(ix) \right]$$

$$\boxed{Y_0(ix) = K_0(x) = -\left(\gamma + \log \frac{x}{2} \right) J_0(x) + \sum_{r=1}^{\infty} \frac{\left(\frac{x}{2} \right)^{2r}}{(r!)^2} \left(1 + \frac{1}{2} + \frac{1}{3} \cdots \frac{1}{r} \right)}$$

at $x = 0$ $\log \frac{x}{2} = \infty$

General soln

$$\boxed{y = A I_0(x) + B K_0(x)}$$

3.10. Bessel and Kelvin functions

3.10.1. ber, bei, ker, kei functions

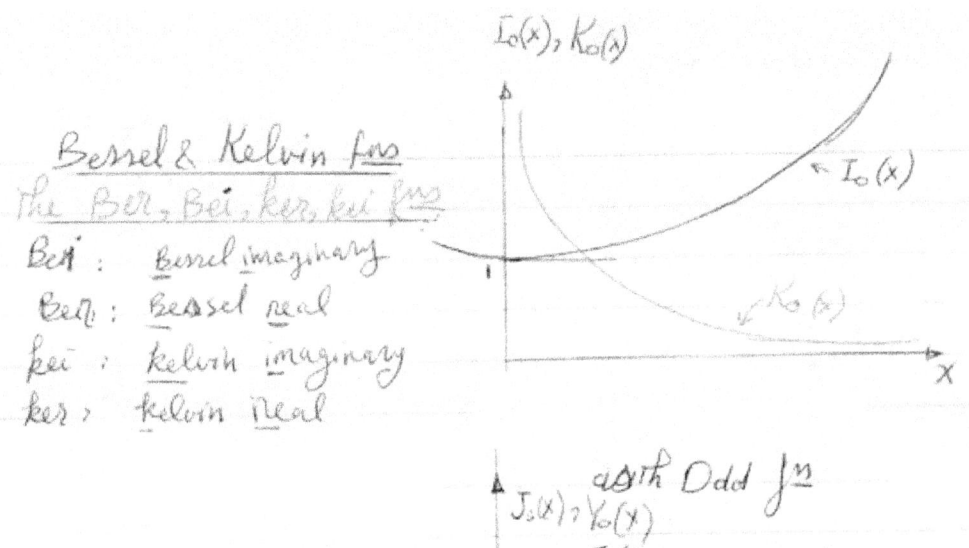

Bessel & Kelvin fns

The Ber, Bei, ker, kei fns

Bei : Bessel imaginary

Ber : Bessel real

kei : kelvin imaginary

ker : kelvin real

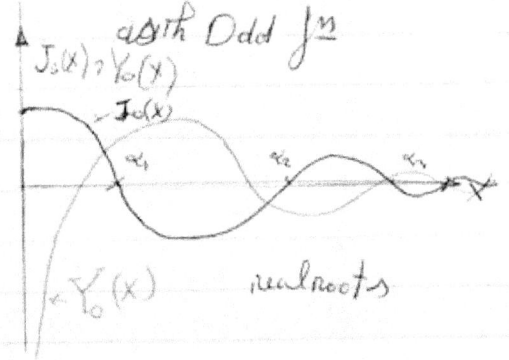

with Odd J^n

real roots

Definition:-

$$I_o (x\sqrt{i}) = ber\, x + i\, bei\, x$$
$$K_o (x\sqrt{i}) = ker\, x + i\, kei\, x.$$

$$I_o(x\sqrt{i}) = 1 + \frac{(x\sqrt{i})^2}{2^2} + \frac{(x\sqrt{i})^4}{2^2 \cdot 4^2} + \frac{(x\sqrt{i})^6}{2^2 \cdot 4^2 \cdot 6^2} + \frac{(x\sqrt{i})^8}{2^2 \cdot 4^2 \cdot 6^2 \cdot 8^2}$$

$$= 1 + \frac{i x^2}{2^2} - \frac{x^4}{2^2 \cdot 4^2} - \frac{i x^6}{2^2 \cdot 4^2 \cdot 6^2} + \frac{x^8}{2^2 \cdot 4^2 \cdot 6^2 \cdot 8^2}$$

$$ber\, x = 1 - \frac{x^4}{2^2 \cdot 4^2} + \frac{x^8}{2^2 \cdot 4^2 \cdot 6^2 \cdot 8^2} - \frac{x^{12}}{2^2 \cdot 4^2 \cdot 6^2 \cdot 8^2 \cdot 10^2 \cdot 12^2}$$

$$bei\, x = \frac{x^2}{2^2} - \frac{x^6}{2^2 \cdot 4^2 \cdot 6^2} + \frac{x^{10}}{2^2 \cdot 4^2 \cdot 6^2 \cdot 8^2 \cdot 10^2} - \frac{x^{14}}{2^2 \cdot 4^2 \cdot 6^2 \cdot 8^2 \cdot 10^2 \cdot 12^2 \cdot 14^2}$$

47

3.11. Bessel functions of any real order

$$25 \backslash 3 \backslash 1972$$

Bessel fns of any real order:-

Case $\textcircled{1}$ n - non-integral رتبة غير صحيحة

$$x^2 \frac{d^2y}{dx^2} + x \frac{dy}{dx} + (x^2 - n^2)y = 0 \qquad \text{dividing by } x^2$$

$$\frac{d^2y}{dx^2} + \frac{1}{x}\frac{dy}{dx} + \left(1 - \frac{n^2}{x^2}\right)y = 0 \underline{\qquad} \textcircled{1}$$

$$y = x^c \left[a_0 + a_1 x + a_2 x^2 + \cdots \right], \quad a_0 \neq 0$$

i.e:

$$y = a_0 x^c + a_1 x^{c+1} + a_2 x^{c+2} + \cdots$$

$$\frac{dy}{dx} = a_0 c x^{c-1} + a_1(c+1)x^c + a_2(c+2)x^{c+1} + \cdots$$

$$\frac{d^2y}{dx^2} = a_0 c(c-1)x^{c-2} + a_1(c+1)c\, x^{c-1} + a_2(c+2)(c+1)x^c + \cdots$$

Substitute from y & its derivatives in $\textcircled{1}$ we get.

$$a_0 c(c-1)x^{c-2} + a_1(c+1)cx^{c-1} + a_2(c+2)(c+1)x^c + \cdots$$

$$a_0 c \quad x^{c-2} + a_1(c+1)x^{c-1} + a_2(c+2)x^c$$

$$- a_0 n^2 x^{c-2} - a_1 n^2 x^{c-1} - a_2 n^2 x^c + \cdots$$

$$+ a_0 x^c + a_1 x^{c+1} + a_2 x^{c+2} \cdots = 0$$

48

<u>coeff^{ts} of x^{c-2}</u> :— $[c^2 - c + c - n^2] a_0 = 0$ ∴ $c^2 - n^2 = 0$

∴ $\boxed{c = \pm n}$ $a_0 \neq 0$

<u>coeff^{ts} of x^{c-1}</u> :— $[c^2 + c + c + 1 - n^2] a_1 = 0$ ∴ $[(c+1)^2 - n^2] a_1 = 0$

$(c+1)^2, n^2 \neq 0$ ∴ $\boxed{a_1 = 0}$

<u>coeff^{t} of x^{c}</u> :—

$$[(c+2)(c+1) + (c+2) - n^2] a_2 + a_0 = 0$$

$$[(c+2)^2 - n^2] a_2 + a_0 = 0 \qquad \boxed{a_2 = \checkmark} \neq 0$$

In general we have. $a_{r+2} [(c+r+2)^c - n^2] + a_r = 0$

$$a_{r+2} = \frac{-a_r}{(c+r+2)^2 - n^2}$$

taking $c = n$ we get

$$a_{r+2} = \frac{-a_r}{n^2 + r^2 + 2^2 + 2nr + 4n + 4r - n^2} = \frac{-a_r}{(r+2)^2 + 2n(r+2)}$$

$$a_{r+2} = -\frac{a_r}{(r+2)(2n+r+2)}$$

Take $r = 0$	$a_2 = -\dfrac{a_0}{2(2n+2)}$	$r = 1$	$a_3 = 0$
			since $a_1 = 0$
$r = 2$	$a_4 = -\dfrac{a_2}{4(2n+4)} = \dfrac{a_0}{4(2n+4)2(2n+2)}$		$a_5 = 0$
			⋮
	$a_6 = \checkmark$		$a_7 = 0$

& hence all coeff^{ts} with (odd sufix) vanish.

$$y = a_0 x^n \left[1 - \frac{x^2}{2(2n+2)} + \frac{x^4}{2 \cdot 4(2n+2)(2n+4)} - \frac{x^6}{2 \cdot 4 \cdot 6(2n+2)(2n+4)(2n+6)} \right.$$

given a_0 its conventional value $y^{...}$ $a_0 = \dfrac{1}{2^n \Gamma(n+1)}$

we obtain $J_n(x)$ which is Bessel fn of the 1st kind of order n. i.e

$$y = J_n(x) = \frac{x^n}{2^n \Gamma(n+1)} \left[1 - \frac{x^2}{2(2n+2)} + \frac{x^4}{2 \cdot 4(2n+2)(2n+4)} + \frac{x^6}{2 \cdot 4 \cdot 6(2n+2)(2n+4)(6+6)} \right]$$

$$J_n(x) = \frac{x^n}{2^n \Gamma(n+1)} \left[1 - \frac{x^2}{2(2n+2)} + \frac{x^4}{2 \cdot 4(2n+2)(2n+4)} - \frac{x^6}{2 \cdot 4 \cdot 6(2n+2)(2n+4)(2n+6)} \right]$$

the 2nd soln is obtained by changing n into $(-n)$ & is denoted by $J_{-n}(x)$ & hence the soln of Bessel eqn is. ✗ ^{1}since $n=$ non integer

$$y = A J_n(x) + B J_{-n}(x)$$

^{1} the denominator will not vanish

linear combination means the 1st soln multiplied by constant A plus + the 2nd soln multiplied by another constant B.

correlation ⁓ Take $n = \frac{1}{2}$ corollary

$$J_{\frac{1}{2}}(x) = \frac{x^{\frac{1}{2}}}{2^{\frac{1}{2}} \cdot \frac{1}{2}\sqrt{\pi}} \left[1 - \frac{x^2}{2 \cdot 3} + \frac{x^4}{2 \cdot 4 \cdot 3 \cdot 5} - \frac{x^6}{2 \cdot 4 \cdot 6 \cdot 3 \cdot 5 \cdot 7} + \frac{x^8}{2 \cdot 4 \cdot 6 \cdot 3 \cdot 3 \cdot 5 \cdot 9} \right]$$

multiply & divide by x

$$J_{\frac{1}{2}}(x) = \sqrt{\frac{2}{\pi x}} \left[x - \frac{x^3}{3!} + \frac{x^5}{5!} - \frac{x^5}{5!} \right]$$

50

$$J_{\frac{1}{2}}(x) = \sqrt{\frac{2}{\pi x}} \sin x .$$

$$J_{\frac{1}{2}}(x) = \sqrt{\frac{2}{\pi x}} \sin x .$$

$$J_{-\frac{1}{2}}(x) = \sqrt{\frac{2}{\pi x}} \cos x .$$

3.12. Bessel functions of integral order

Case (I) n is an integer or zero

The 1st sol^n is the same as before. Since when (n) is a positive integer
$\therefore \Gamma(n+1) = n!$. Then $J_n(x)$ can be rewritten as the form.

$$J_n(x) = \frac{x^n}{2^n n!} \left[1 - \frac{x^2}{2(2n+2)} + \frac{x^4}{2.4(2n+2)(2n+4)} - \cdots \right]$$

The 2nd sol^n cannot be obtained by changing n to $-n$ as in the previous case. Since some of the denominators become zero.

The standard 2nd sol^n is given by.

$$Y_n(x) = \frac{2}{\pi} \left\{ \gamma + \log \frac{x}{2} \right\} J_n(x) - \frac{1}{\pi} \sum_{r=0}^{n-1} \frac{(n-r-1)!}{r!} \left(\frac{2}{x}\right)^{n-2r}$$

$$- \frac{1}{\pi} \sum_{r=0}^{\infty} \frac{(-1)^r (\frac{x}{2})^{2r+n}}{r! (n+r)!} \left\{ 1 + \frac{1}{2} + \frac{1}{3} + \cdots + \frac{1}{r} + 1 + \frac{1}{2} + \frac{1}{3} + \cdots \right.$$

$$\left. + \cdots + \frac{1}{n+r} \right\}$$

The 2nd sol^n should not be applied in any region containing the origin since the f^n becomes ∞ when $x=0$ due to the presence of the logarithmic term.

Bessel coefficients:-
$$e^{\frac{x}{2}(t - \frac{1}{t})} = \sum_{n=-\infty}^{\infty} J_n(x) t^n \qquad t \neq 0$$

3.13. Bessel coefficients

i.e: $J_n(x)$ is the coefficient of (t^n) in the expansion of the exponential f.s which is called the generating function.

$L.H.S = e^{\frac{xt}{2}} \cdot e^{-\frac{x}{2t}}$

Now

$$e^x = 1 + x + \frac{x^2}{2!} + \cdots + \frac{x^n}{n!} + \cdots$$

$$e^{\frac{xt}{2}} = 1 + \frac{xt}{2} + \frac{1}{2!}\left(\frac{xt}{2}\right)^2 + \cdots + \frac{1}{n!}\left(\frac{xt}{2}\right)^n + \frac{1}{(n+1)!}\left(\frac{xt}{2}\right)^{n+1} + \cdots$$

$$e^{-\frac{x}{2t}} = 1 - \frac{x}{2t} + \frac{1}{2!}\left(-\frac{x}{2t}\right)^2 + \cdots + \frac{1}{n!}\left(-\frac{x}{2t}\right)^n + \frac{1}{(n+1)!}\left(\frac{-x}{2t}\right)^{n+1} + \cdots$$

$$e^x = 1 + \frac{x}{1!} + \frac{x^2}{2!} + \frac{x^3}{3!} + \cdots + \frac{x^n}{n!} + \cdots$$

$$e^{\frac{xt}{2}} = 1 + \frac{xt}{2} + \frac{1}{2!}\left(\frac{xt}{2}\right)^2 + \cdots + \frac{1}{n!}\left(\frac{xt}{2}\right)^n + \frac{1}{(n+1)!}\left(\frac{xt}{2}\right)^{n+1} + \cdots$$

$$e^{-\frac{x}{2t}} = 1 - \frac{x}{2t} + \frac{1}{2!}\left(\frac{x}{2t}\right)^2 + \cdots + \frac{1}{n!}\left(\frac{-x}{2t}\right)^n + \frac{1}{(n+1)!}\left(\frac{-x}{2t}\right)^{n+1} + \cdots$$

<u>coeffts of t^n in the product.</u>

$$\frac{x^n}{2^n\, n!} - \frac{x^{n+2}}{2^{n+2}(n+1)!} + \frac{x^{n+4}}{2^{n+6}\,(n+2)!} \cdots \qquad 2.4\,(2n+2)(2n+4)$$

i.e. $\dfrac{x^n}{2^n\, n!}\left\{1 - \dfrac{x^2}{2^2(n+1)} + \dfrac{x^4}{2^6(n+2)(n+1)} - \cdots\right\} = J_n(x)$

which is $J_n(x)$ since J is the coefft of t^n. next consider <u>the coefft of $t^{-n} = \frac{1}{t^n}$</u> this is given by

$$\frac{(-1)^n x^n}{2^n n!} + \frac{(-1)^{n+1} x^{n+2}}{2^{n+2}(n+1)!} + \cdots$$

i.e. $\quad = \frac{(-1)^n x^n}{2^n n!} \left[1 - \frac{x^2}{2(2n+2)} + \frac{x^4}{2.4(2n+2)(2n+4)} - \cdots \right] = J_{-n}(x)$

$\underset{2(n+1)}{h}$

Since it is the coefficient of t^{-n}. $J_{-n}(x)$ which is a <u>Bessel</u> coeff<u>t</u> being the colfft of t^{-n} in the expansion of the generating fn (generating fn). we notice that the cofft. (Bessel cofts)

$$J_{-n}(x) = (-1)^n J_n(x)$$

3.14. Recurrence formulae

N/4/1972

Recurrence Formulæ صيغ

$$e^{\frac{x}{2}(t - \frac{1}{t})} = \sum_{r=-\infty}^{\infty} J_n(x) t^n. \quad \text{— (1)}$$

Differentiate (1) w.r.t. (t)

$$\frac{x}{2}(1 + t^{-2})(e^{\frac{x}{2}(t - \frac{1}{t})}) = \sum_{r=-\infty}^{\infty} n J_n(x) t^{n-1}$$

$$\frac{x}{2}(1 + t^{-2}) \sum_{n=-\infty}^{\infty} J_n(x) t^n = \sum_{n=-\infty}^{\infty} n J_n(x) t^{n-1}$$

53

Equating the coefft's of t^{n-1} in both sides we get.

$$\frac{x}{2} J_{n+1}(x) + \frac{x}{2} J_{n-1}(x) = n J_n(x)$$

$$\boxed{J_{n-1}(x) + J_{n+1}(x) = \frac{2n}{x} J_n(x)} \quad \text{—②}$$

This is a recurrence formula connecting every 3 - Bessel J_n

of consecutive order.

again differentiate ① w.r.t Ⓧ

$$\frac{1}{2}(t - t^{-1}) \sum_{n=-\infty}^{\infty} J_n(x) t^n = \sum_{n=-\infty}^{\infty} J_n'(x) t^n$$

Equating the coefft in both sides we get, t^n we get

$$\frac{1}{2} J_{n-1} + \frac{1}{2} J_{n+1} = J_n'$$

$$\boxed{J_{n-1}(x) - J_{n+1}(x) = 2 J_n'(x)} \quad ③$$

Adding ② + ③ : $J_{n-1} = \frac{n}{x} J_n + J_n'$ multiply both sides by x^n

$$\therefore x^n J_{n-1} = n x^{n-1} J_n + x^n J_n'$$

$$\boxed{x^n J_{n-1} = \frac{d}{dx}\left[x^n J_n\right]} \quad \text{—④}$$

or $$\boxed{\int x^n J_{n-1}(x)\, dx = \frac{n}{x} J_n(x)} \quad \text{—⑤}$$

Subtracting ③ from ② we get $J_{n+1} = \frac{n}{x} J_n - J_n'$ multiply both sides by $-x^{-n}$

54

$$\therefore \quad -x^{-n} J_{n+1} = -nx^{-n-1} J_n + x^{-n} J_n'$$

$$\therefore \boxed{-x^{-n} J_{n+1}(x) = \frac{d}{dx}\left[x^{-n} J_n(x) \right]} \longrightarrow \text{⑥}$$

$$\text{or} \quad \boxed{\int \frac{J_{n+1}}{x^n} \, dx = -\frac{J_n(x)}{x^n}} \longrightarrow \text{⑦}$$

3.15. Important expansions in Bessel functions

Some important expansions

$$e^{\frac{x}{2}(t-\frac{1}{t})} = \sum_{n=-\infty}^{\infty} J_n(x) t^n = J_0(x) + J_1(x)t + J_2(x)t^2 + \qquad n = 0 \to \infty, \, 0 \to -\infty$$
$$+ J_{-1}(x) t^{-1} + J_{-2}(x) t^{-2} + \cdots$$

put $t = e^{i\theta}$

$$\therefore e^{\frac{x}{2}(e^{i\theta} - e^{-i\theta})} = J_0(x) + J_1(x) e^{i\theta} + J_2(x) e^{2i\theta} + \cdots$$
$$+ J_{-1}(x) e^{-i\theta} + J_2(x) e^{-2i\theta} + \cdots$$

$$\therefore \quad J_{-n}(x) = (-1)^n J_n(x)$$

$$\therefore \quad J_{-1}(x) = -J_1(x)$$

$$\therefore \quad J_{-2}(x) = J_2(x)$$

$$\therefore \quad \cos\theta = \frac{e^{i\theta} + e^{-i\theta}}{2} \, , \quad \sin\theta = \frac{e^{i\theta} - e^{-i\theta}}{2i}$$

$$\therefore e^{ix\sin\theta} = J_0(x) + J_1(x)\left\{ e^{i\theta} - e^{-i\theta} \right\} + J_2(x)\left\{ e^{2i\theta} + e^{-2i\theta} \right\} + \cdots$$

$\therefore \quad \cos(x\sin\theta) + i\sin(x\sin\theta) = J_0(x) + 2J_1(x)\,i\sin\theta$
$$+ 2J_2(x)\cos 2\theta + \cdots$$

Equating real to real & imaginary to imaginary

$$\cos(x\sin\theta) = J_0(x) + 2\left\{ J_2(x)\cos 2\theta + J_4(x)\cos 4\theta + \cdots \right\}$$

$$\cos(x\sin\theta) = J_0(x) + 2\sum_{P=1}^{\infty} J_{2P}(x)\cos 2P\theta$$

odd. زوجي

$$\sin(x\sin\theta) = 2\left\{ J_1(x)\sin\theta + J_3(x)\sin 3\theta + \cdots \right\}$$

$$= 2\sum_{P=1}^{\infty} J_{2P-1}(x)\sin(2P-1)\theta$$

even.

In the above formulii change θ into $\left(\frac{\pi}{2} - \theta\right)$

$$\cos(x\cos\theta) = J_0(x) + 2\left\{ - J_2(x)\cos 2\theta + J_4(x)\cos 4\theta \right.$$

$$= J_0(x) - 2\left\{ J_2(x)\cos 2\theta - J_4(x)\cos 4\theta \right.$$

$$\sin(x\cos\theta) = 2\left\{ J_1(x)\cos\theta - J_3(x)\cos(3\theta) + \cdots \right.$$

now put $\theta = 0$

$$\cos x = J_0(x) - 2\left\{ J_2(x) - J_4(x) + J_6(x) \cdots \right\}$$

$$\sin x = 2\left\{ J_1(x) - J_3(x) + J_5(x) + \cdots \right\}$$

the eqn shows the diffinite relationship betn circular fn & bessel fn of 1st kind

$$\cos x = J_0(x) - 2 \left\{ J_2(x) - J_4(x) + J_6(x) \cdots \right\}$$

$$\sin x = 2 \left\{ J_1(x) - J_3(x) + J_5(x) - \cdots \right\}.$$

3.16. Bessel function as integrals

Bessel fn as integrals

$$e^{\frac{x}{2}\left(t - \frac{1}{t}\right)} = \sum_{n=-\infty}^{\infty} J_n(x) t^n \qquad \text{put } t = e^{i\theta}$$

$$e^{ix\sin\theta} = \sum_{n=-\infty}^{\infty} J_n(x) e^{in\theta} \qquad \text{chang } n \to n+r$$

$$e^{ix\sin\theta} = \sum_{r=-\infty}^{\infty} J_{n+r}(x) e^{i(n+r)\theta}$$

$$= e^{in\theta} \sum_{r=-\infty}^{\infty} J_{n+r}(x) e^{ir\theta}$$

$$e^{-i(n\theta - x\sin\theta)} = \sum_{r=-\infty}^{\infty} J_{n+r}(x) e^{ir\theta}$$

from Euler exp.

$$\cos(n\theta - x\sin\theta) - i\sin(n\theta - x\sin\theta) = \sum_{r=-\infty}^{\infty} J_{n+r}(x) e^{ir\theta}$$

integral w.r.t. (θ) lets $\theta = 0$, $\theta = 2\pi$

$$\int_0^{2\pi} e^{ir\theta} d\theta = \left[\frac{\sin(r\theta) - i\cos(r\theta)}{r} \right]_0^{2\pi} = 0 \quad r \neq 0$$

57

when $r=0$ is $\int_0^{2\pi} e^{ir\theta} \, d\theta = 2\pi$

is $\int_0^{2\pi} \{\cos(n\theta - x\sin\theta) \, d\theta = 2\pi \, J_n(x)$

$$\boxed{J_n(x) = \frac{1}{2\pi} \int_0^{2\pi} \cos[n\theta - x\sin\theta] \, d\theta}$$

n is a +ve integer.

3.17. The Bessel functions of order n of the third kind (Hankel functions of order n)

Bessel f^n of the 3rd kind (Hankel f^{ns}) german

In many practical problems it is more convenient to use complex combinations of Bessel f^{ns} of 1st or 2nd kind. These are

$$H_n^{(1)}(x) = J_n(x) + i \, Y_n(x)$$

$$H_n^{(2)}(x) = J_n(x) - i \, Y_n(x)$$

$H_n^{(1)}(x)$, $H_n^{(2)}(x)$ are known as Hankel f^{ns}

4. Legendre functions

(Correction: the handwritten error "Lagendre" is corrected as "Legendre".)

Legendre Functions

there are sol^{n}s of Legendre diff^l eq^n.

$$(1-x^2)\frac{d^2y}{dx^2} - 2x\frac{dy}{dx} + n(n+1)y = 0 \qquad \text{———①}$$

n is a constant but from a practical pt of view the important case is when $\underline{n}$ is a positive integer : it is more convenient to obtain the sol^n of the eq^n in <u>des-cending</u> power of x

let

$$y = a_0 x^m + a_1 x^{m-1} + a_2 x^{m-2} + \ldots \qquad \text{L} \; a \neq 0$$

$$\frac{dy}{dx} = a_0 m x^{m-1} + a_1(m-1)x^{m-2} + a_2(m-2)x^{m-3} +$$

$$\frac{d^2y}{dx^2} = a_0 m(m-1)x^{m-2} + a_1(m-1)(m-2)x^{m-3} + a_2(m-2)(m-3)x^{m-4} \ldots$$

substituting in ①

$$a_0 m(m-1)x^{m-2} + a_1(m-1)(m-2)x^{m-3} + \ldots$$
$$- a_0 m(m-1)x^m - a_1(m-1)(m-2)x^{m-1} - a_2(m-2)(m-3)x^{m-2} + \ldots$$
$$- 2a_0 m x^m - 2a_1(m-1)x^{m-1} - 2a_2(m-2)x^{m-2} + \ldots$$
$$+ n(n+1)a_0 x^m + n(n+1)a_1 x^{m-1} + n(n+1)a_2 x^{m-2} + \ldots = 0$$

<u>coeff^{t}s of x^m</u>

$$- a_0\left[m(m-1) + 2m - n(n+1)\right] = 0$$

as $a_0 \neq 0$ ∴ $m^2 - m + 2m - n^2 - n = 0$

$$(m-n)(m+n+1) = 0$$

$$\boxed{m = n \text{ or } -(n+1)} \qquad \text{———②}$$

II $\boxed{1^{st} \text{ sol}^n \quad m = n}$

$\underline{\text{colff}^t \text{ y } x^{m-1}}$

$$- a_1 \left[(m-1)(m-2) + 2(m-1) - n(n+1) \right] = 0$$

putting $m = n$ we get, ———②

$$a_1 \left[n^2 + 2 - 3n + 2n - 2 - n^2 - n \right]$$

$$- 2 a_1 n = 0$$

$$\therefore \boxed{a_1 = 0}$$

$\underline{\text{colff}^t \text{ y } x^{m-2}}$

$$a_0 \, m(m-1) - a_2 \left[(m-2)(m-3) + 2(m-2) - n(n+1) \right]$$

putting $m = n$. ———③

$$a_0 \, n(n-1) - a_2 \left[n^2 - 5n + 6 + 2n - 4 - n^2 - n \right] = 0$$

$$a_0 \, n(n-1) - a_2 \left[2 - 4n \right] = 0$$

$$\boxed{a_2 = \frac{-n(n-1)}{2(2n-1)} \; a_0} \quad ———$$

similarly $\quad a_0 \neq 0 \qquad\qquad\qquad a_1 = 0$

$$a_2 = \frac{-n(n-1)}{2(2n-1)} \, a_0 \qquad\qquad a_3 = 0$$

$$a_4 = \frac{n(n-1)(n-2)(n-3)}{2.4(2n-1)(2n-3)} \qquad a_5 = 0$$

$$a_6 = \frac{-n(n-1)(n-2)(n-3)(n-4)(n-5)}{2.4.6(2n-1)(2n-3)(2n-5)} \qquad a_7 = 0$$

$$\therefore \quad y = a_0 \left[x^n - \frac{n(n-1)}{2(2n-1)} x^{n-2} + \frac{n(n-1)(n-2)(n-3)}{2.4(2n-1)(2n-3)} x^{n-4} - \cdots \right]$$

60

giving a_0 its conventional value namely.

$$a_0 = \frac{(2n)!}{2^n (n!)^2}$$

we obtain $P_n(x)$ i.e.

1st sol^n

$$P_n(x) = \frac{(2n)!}{2^n (n!)^2} \left[x^n - \frac{n(n-1)}{2(2n-1)} x^{n-2} + \frac{n(n-1)(n-2)(n-3)}{2 \cdot 4 (2n-1)(2n-3)} x^{n-4} - \right]$$

The series terminates & $P_n(x)$ is called the gendre polynomial in many practical problems x stands for $\cos\theta$ by giving n the values $0, 1, 2, 3 \ldots$ we get $n=4$

$$P_4(x) = \frac{8!}{2^4 (4!)^2} \left[x^4 - \frac{4 \times 3}{2 \times 7} x^2 + \frac{4 \times 3 \times 2 \times 1}{2 \cdot 4 \cdot 7 \cdot 5} \right]$$

$$P_4(x) = \frac{35}{8} \left(x^4 - \frac{6}{7} x^2 + \frac{3}{35} \right) = \frac{1}{8} (35x^4 - 30x^2 + 3)$$

Note   sum of coeffts $\frac{35 - 30 + 3}{8} = 1$ and always   (1)

$P_0(x) = 1$ $P_3(x) = \frac{1}{2}(5x^3 - 3x)$

$P_1(x) = x$ $P_4(x) = \frac{1}{8}(35x^4 - 30x^2 + 3)$

$P_2(x) = \frac{1}{2}(3x^2 - 1)$

The 2nd sol^n is obtained in a similar way by putting

$m = -(n+1)$

61

$$\boxed{\text{II} \quad 2\text{nd Sol}^{\underline{n}} \quad m = -(n+1)} \qquad 2\text{nd sol}^{\underline{n}}$$

$$Q_n(x) = \left(\frac{2^n (n!)^2}{(2n+1)!}\right)\left[x^{-n-1} + \frac{(n+1)(n+2)}{2(2n+3)} x^{-n-3} + \frac{}{} \right.$$

$$\text{y-line} \qquad \left. + \frac{(n+1)(n+2)(n+3)(n+4)}{2\cdot 4(2n+3)(2n+5)} x^{-n-5} + \cdots \right]$$

Here the series does not terminate & series is convergent if

modulus $|x| > 1$ $\quad \frac{1}{x^{n+1}} \quad$ indefinite

$= \infty$ at origin like $\log x$ —

proof $\quad$ in (?) $\quad$ put $m = -(n+1)$. $\qquad$ biology

coefft of x^{m-1} $\qquad$ Bessel

$$a_1 \left[(-n-2)(-n-3) + 2(-n-2) - n(n+1)\right] = 0$$

$$a_1 \left\{n^2 + 5n + 6 - 2n - 4 - n^2 - n\right\} = 0$$

$$a_1 \left[2n + 2\right] \qquad \therefore \boxed{a_1 = 0}$$

coefft of x^{m-2}

$$- a_0(n+1)(-n-2) - a_2\left\{(-n-3)(-n-4) + 2(-n-3) - n(n+1)\right\}$$

$$+ a_0(n+1)(n+2) - a_2\left[n^2 + 7n + 12 - 2n - 6 - n^2 - n\right] = 0$$

$$a_0(n+1)(n+2) - a_2\left[4n + 6\right] = 0$$

$$\boxed{a_2 = \frac{(n+1)(n+2)}{2(2n+3)} a_0}$$

$$a_0 \neq 0 \qquad\qquad a_1 = 0$$

$$a_2 = \frac{(n+1)(n+2)}{2(2n+3)} a_0 \qquad a_3 = 0$$

$$a_4 = \frac{(n+1)(n+2)(n+3)(n+4)}{2\cdot 4(2n+3)(2n+5)} a_0 \qquad a_5 = 0$$

4.1. Alternative definition of Legendre polynomials

The general solution is

$$y = A P_n(x) + B Q_n(x)$$

Alternative defn. of Legendre polynomials.

$$\phi = (1 - 2xh + h^2)^{-\frac{1}{2}} = \sum_{n=0}^{\infty} P_n(x) h^n \quad \text{——①}$$

$$= P_0(x) + P_1(x)h + P_2(x)h^2 + \cdots + P_n(x)h^n + \cdots$$

i.e : $P_n(x)$ is the coefft. of h^n in the expansion of ϕ in ascending +ve powers of h.

$$\phi = [1 - (2xh - h^2)]^{-\frac{1}{2}}$$

$$= 1 + \frac{1}{2}(2xh - h^2) - \frac{1}{2} \times \frac{3}{2} \times \frac{1}{2!}(2xh - h^2)^2 - \cdots$$

$$=$$

$$= 1 + xh + \frac{1}{2}(3x^2 - 1)h^2 + \cdots$$

$$\phi = P_0(x) + P_1(x)h + P_2(x)h^2 + \cdots$$

To obtain the Recurrence formullä we diff. partially w.r.t h

4.2. Legendre's recurrence formulae

Recurrence formullä

Differential partially w.r.t h

$$\frac{\partial \phi}{\partial h} = -\frac{1}{2}(1 - 2xh + h^2)^{-\frac{3}{2}}(-2x + 2h)$$

$$\frac{\partial \phi}{\partial h} = \frac{(x - h)\phi}{1 - 2xh + h^2}.$$

$$\therefore (x - h)\phi = (1 - 2xh + h^2)\frac{\partial \phi}{\partial h}$$

$$(x - h)\phi \equiv (1 - 2xh + h^2)\frac{\partial\phi}{\partial h}$$

$$= (x-h)(P_0 + P_1 h + \cdots + P_n h^n + \cdots]\cdots$$

$$\equiv (1 - 2xh + h^2)(P_1 + 2P_2 h + \cdots + nP_n h^{n-1} + \cdots]$$

equating the coeff^{ts} of h^n on both sides we get.

$$x P_n - P_{n-1} = (n+1)P_{n+1} - 2nx P_n + (n-1)P_{(n-1)}$$

arrange.

$$\therefore \boxed{(n+1) P_{n+1} - (2n+1)x P_n + n P_{(n-1)} = 0} \quad \boxed{I}$$

this is reccurrent relations connecting three of Legendre polynomials of conseautive degrees.

<u>again   Differentiat w.r.t (x)</u>

$$\phi = (1 - 2xh + h^2)^{-\frac{1}{2}} = \sum_{n=0}^{\infty} P_n(x) h^n \quad \text{---} \quad \textcircled{1}$$

$$\frac{\partial\phi}{\partial x} = -\frac{1}{2}(1 - 2xh + h^2)^{-\frac{3}{2}} \cdot (-2h)$$

$$\frac{\partial\phi}{\partial x} = \frac{h\phi}{1 - 2xh + h^2}$$

$$h\phi = (1 - 2xh + h^2)\frac{\partial\phi}{\partial x}$$

64

$$h(\phi) = (1 - 2xh + h^2)\frac{\partial \phi}{\partial x}$$

$$h(P_0 + P_1 h + \cdots + P_n h^n + \cdots) = (1 - 2xh + h^2)(P_1' h + P_2' h^2 \cdots P_n' h^n + \cdots)$$

dashes mean differentiating w.r.t x.
equating coeff$\underline{ts}$ of h^{n+1} we get.

$$\boxed{P_n = P_{(n+1)}' - 2x P_n' + P_{(n-1)}'} - \textcircled{II}$$

important

we can now from ① prove the property that the sum of the coefficients in any legendre polynomial is unity .. in ① put $x = 1$

$$\phi = (1 - 2h + h^2)^{-\frac{1}{2}} = [(1-h)^2]^{-\frac{1}{2}} = \frac{1}{1-h}$$

$$= 1 + h + h^2 + \cdots + h^n + = \sum_{n=0}^{\infty} P_n(1) h^n$$
$$= P_0 + P_1 h + P_2 h^2 + \cdots + P_n h^n$$

equating coff$\underline{ts}$ of h^n we get

$$P_n(1) = 1 \qquad \text{that is the sum of the coff}\underline{ts} = \underline{1} \text{ unity}$$

$\therefore$ $(n+1)P_{n+1} - (2n+1)x P_n + n P_{n-1} = 0 \quad\text{———} \textcircled{1}$

$\qquad P_n = P_{n+1}' - 2x P_n' + P_{n-1}' \quad\text{———} \textcircled{2}$

$\therefore \dfrac{\partial \phi}{\partial h} \Big/ \dfrac{\partial \phi}{\partial x} = \dfrac{(x-h)\phi}{h\phi}$

$\therefore \therefore (x-h)\dfrac{\partial \phi}{\partial x} = h\dfrac{\partial \phi}{\partial h}$

$$\Rightarrow : (x-h)(P'_0(x) + P'_1(x)h + P'_2 h^2 + \cdots P'_n h^n)$$
$$= h(P_1 + 2P_2 h + 3P_3 h^2 + \cdots + nP_n h^{n-1})^n \quad \boxed{III}$$

equating h^n coeffts

$$xP'_n - P'_{n-1} = (n+1)P_{n+1} \qquad \boxed{xP'_n - P'_{n-1} = nP_n}$$

multiply III by 2 & add to ② eqn

$$P_n = P'_{n+1} - 2xP'_n + P'_{n-1} \qquad \boxed{IV}$$

$$2xP'_n - 2P'_{n-1} = 2nP_n \qquad \boxed{(2n+1)P_n = P'_{n+1} - P'_{n-1} \ \textcircled{c}}$$

4.3. Integral properties of Lagendre polynomial

Integral property of Lagendre polynomial.

It can be shown that.

$$\int_{-1}^{+1} P_m(x) \cdot P_n(x)\, dx = 0 \qquad \text{if } m \neq n$$
$$= \frac{2}{2n+1} \quad \text{if } m = n$$

we can now use this integral properties in expanding the given f^n $F(x)$ in a series of Lagendre polynomials.

$$F(x) = a_0 P_0(x) + a_1 P_1(x) + \cdots + a_n P_n(x) + \cdots$$

to obtain a_n multiply both sides by the cffft of a_n namely $P_n(x)$ & then integrate bet^n $-1, 1$

$$\therefore \int_{-1}^{1} F(x) P_n(x)\, dx = a_n \int_{-1}^{1} \{P_n(x)\}^2 dx = \frac{a_n \cdot 2}{2n+1}$$

$$\therefore a_n = \frac{2n+1}{2} \int_{-1}^{1} F(x) P_n(x)\, dx.$$

66

4.4. The associated Legendre functions

15 \ 9 \ 1972

The associated Legendre fns:—

These are $\frac{\text{soln}^s}{}$ of the associated Legendre diff^l eq^n

$$(1-x^2)\frac{d^2y}{dx^2} - 2x\frac{dy}{dx} + \left\{ n(n+1) - \frac{m^2}{1-x^2} \right\} y = 0$$

m, n are constants but from a practical p^t of view the important case is when m & n are $+ve$ integers & $m \leq n$ the soln^s are.

$$P_n^m(x) = (1-x^2)^{\frac{m}{2}} \frac{d^m}{dx^m} \left\{ P_n(x) \right\} \quad — ① \quad \text{associated Legendre function } (يعني) \quad m \text{ not exceed } n$$

$$Q_n^m(x) = (1-x^2)^{\frac{m}{2}} \frac{d^m}{dx^m} \left\{ Q_n(x) \right\} \quad — ②$$

4.5. <u>APPLICATION</u>S OF LEGENDRE FUNCTIONS

4.5.1. Laplace's wave equation in spherical polar coordinates

4.5.1.1. <u>Application</u> to heat transfer

Applications.

problem:— A solid sphere of radius unity. has one half of its surface kept at constant temp. $1°C$. while the other surface kept at the constant temp. $0°C$. when the flow of heat is steady find the permilant temp at p^t in solid sphere.

Solution.

when the flow of heat is steady the temp. at pt. inside the sphere U is satisfies Laplace's eq^n.

$$\frac{\partial}{\partial r}\left(r^2 \frac{\partial U}{\partial r}\right) + \frac{1}{\sin\theta}\frac{\partial}{\partial \theta}\left(\sin\theta \frac{\partial U}{\partial \theta}\right) + \frac{1}{\sin^2\theta}\frac{\partial^2 U}{\partial \phi^2} = 0 - (1)$$

from orthogonal curvilinear coordinates.

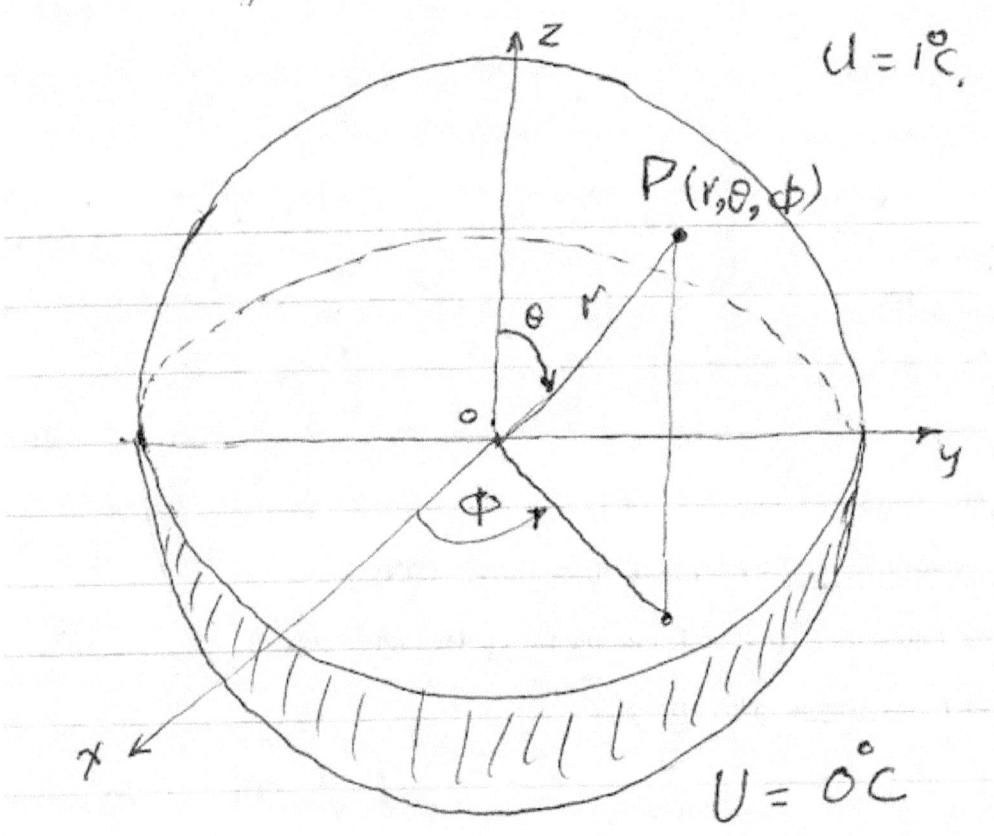

$U = 1°C,$

$P(r, \theta, \phi)$

$U = 0°C$

Here we have symmetry about the z axis & hence U is indept of ϕ & laplaces eq^n reduces to

$$\frac{\partial}{\partial r}\left(r^2\frac{\partial U}{\partial r}\right) + \frac{1}{\sin\theta}\frac{\partial}{\partial\theta}\left(\sin\theta\frac{\partial U}{\partial\theta}\right) = 0 \quad\text{——②}$$

put $U = R\,\Theta$ where R is fn of (r) only & Θ is a fn in θ only.   substituting in ②

$$\frac{\partial}{\partial r}\left(r^2\frac{dR}{dr}\Theta\right) + \frac{1}{\sin\theta}\frac{\partial}{\partial\theta}\left(\sin\theta\frac{d\Theta}{d\theta}\cdot R\right) = 0 \quad\text{——③}$$

dividing by $R\Theta$

$$\therefore \frac{1}{R}\frac{d}{dr}\left(r^2\frac{dR}{dr}\right) + \frac{1}{\Theta\sin\theta}\frac{d}{d\theta}\left(\sin\theta\frac{d\Theta}{d\theta}\right) = 0 \quad\text{——④}$$

$$\therefore \frac{1}{R}\left(r^2\frac{d^2R}{dr^2} + 2r\frac{dR}{dr}\right) = n(n+1) \quad \text{say}$$

$$\frac{1}{\Theta\sin\theta}\frac{d}{d\theta}\left(\sin\theta\frac{d\Theta}{d\theta}\right) = -(n+1) \qquad\Bigg\}\text{——⑤}$$

Hence we have.

$$r^2\frac{d^2R}{dr^2} + 2r\frac{dR}{dr} - n(n+1)R = 0 \quad\text{——①}$$

$$\frac{1}{\sin\theta}\frac{d}{d\theta}\left(\sin\theta\frac{d\Theta}{d\theta}\right) + n(n+1)\Theta = 0 \quad\text{——②}$$

Consider eq^n ① . to solve this eq^n we make use of the R.H.S to be equal $= 0$ & put $\boxed{R = Ar^\lambda.}$

$$\therefore \quad r^2 \cdot A\lambda(\lambda-1)r^{\lambda-2} + 2r \cdot A\lambda r^{\lambda-1} - n(n+1)Ar^\lambda = 0$$

$$\therefore \quad \lambda(\lambda-1)r^\lambda + 2\lambda r^\lambda - n(n+1)r^\lambda = 0$$

$$\therefore \quad \lambda(\lambda-1) + 2\lambda - n(n+1) = 0$$

$$\therefore \quad \lambda^2 + \lambda - n^2 - n = 0$$

$$(\lambda-n)(\lambda+n+1) = 0$$

$$\boxed{\lambda = n \ \text{or} \ -(n+1)}$$

$$\therefore \quad \boxed{R = Ar^n + \frac{B}{r^{n+1}}}$$

to solve eq^n (II) we write this eq^n in the form

$$\frac{1}{-\sin\theta}\frac{d}{d\theta}\left(\sin^2\theta \frac{d\Theta}{-\sin\theta d\theta}\right) + n(n+1)\Theta = 0$$

put $x = \cos\theta$, $dx = -\sin\theta \, d\theta$

$$\therefore \quad \frac{d}{dx}\left[(1-x^2)\frac{d\Theta}{dx}\right] + n(n+1)\Theta = 0$$

$$\boxed{(1-x^2)\frac{d^2\Theta}{dx^2} - 2x\frac{d\Theta}{dx} + n(n+1)\Theta = 0}$$

Which is Lagendre diff^l eq^n. One form of s^{o}l^n is $P_n(x)$
i.e $P_n(\cos\theta) = \Theta$

famous

$$\therefore \quad \boxed{U = R\Theta = \left(Ar^n + \frac{B}{r^{n+1}}\right)P_n(\cos\theta)}$$

Since U is finite at the origine $(r=0)$ then $B=0$ —— ①

&
$$U = Ar^n . P_n(\cos\theta)$$

This is a solut^n for various integral values of n hence a linear combination of this is also a sol^n the diff^l eq^n being linear

$$U = \sum A_m r^n P_n(\cos\theta)$$

when $r=1$, $U = U_s$ is the temp on the surface i.e

$U_s = 0$ at $\frac{\pi}{2} < \theta < \pi$

$U_s = 1$ at $0 < \theta < \frac{\pi}{2}$

$r=1$

$\therefore U_s = \sum A_n P_n \cos(\theta)$

$$U_s = A_0 P_0(\cos\theta) + A_1 P_1(\cos\theta) + \cdots + A_n P_n(\cos\theta).$$

again put $\cos\theta = x$
$$U_s = A_0 P_0(x) + A_1 P_1(x) + \cdots + A_n P_n(x).$$

where

$U_s = 1$ $0 < x < 1$

$U_s = 0$ $-1 < x < 0$

$$\int_{-1}^{1} U_s P_n(x)\,dx = A_n \int_{-1}^{1} \left\{P_n(x)\right\}^2 dx$$

$0 \to$ increas

$\cos\theta \to$ decreases

$$= A_n \frac{2}{2n+1}$$

$$\therefore A_n = \frac{2n+1}{2} \int_{-1}^{1} U_s \, P_n(x) \, dx$$

$$= \frac{2n+1}{2} \int_{0}^{1} 1 \times P_n(x) \, dx \qquad \text{becaus } L.H.S \\ \text{of curve} = 0$$

$\underline{n=0}$
$$A_0 = \frac{1}{2} \int_{0}^{1} P_0(x) \, dx = \frac{1}{2} \int_{0}^{1} dx = \frac{1}{2}$$

$\underline{n=1}$
$$A_1 = \frac{3}{2} \int_{0}^{1} P_1(x) \, dx = \frac{3}{2} \int_{0}^{1} x \, dx = \frac{3}{4}$$

& similarly for the other coefficients, hence we get:

$$U_s = \frac{1}{2} + \frac{3}{4} P_1(x) - \frac{7}{16} P_3(x) \cdots$$

i.e. $U_s = \frac{1}{2} + \frac{3}{4} P_1(\cos\theta) - \frac{7}{16} P_3(\cos\theta) + \cdots$

$$\therefore U(r,\theta) = \frac{1}{2} + \frac{3}{4} r P_1(\cos\theta) - \frac{7}{16} r^3 P_3(\cos\theta) + \cdots$$

4.5.1.2. <u>Application</u> to gravitational potential

Example:- Find the gravitational Potential due to a thin homogeneous circular disc of surface density σ & radius C. **soln**

It is known that at pt devoid of attracting matter the potential satisfies Laplace's eqn.

Laplaces eqn in spherical polar coordinates:-

$$\frac{\partial}{\partial r}\left(r^2 \frac{\partial V}{\partial r}\right) + \frac{1}{\sin\theta}\frac{\partial}{\partial\theta}\left(\sin\theta \frac{\partial V}{\partial\theta}\right) + \frac{1}{\sin^2\theta}\frac{\partial^2 V}{\partial\phi^2} = 0$$

here we have symmetry about z axis i.e:- potential V is independent on ϕ. Laplac's eqn reduces to

$$\frac{\partial}{\partial r}\left(r^2 \frac{\partial V}{\partial r}\right) + \frac{1}{\sin\theta}\frac{\partial}{\partial\theta}\left(\sin\theta \frac{\partial V}{\partial\theta}\right) = 0$$

as in the previous problem we have a type of solns.

$$\sum A_n r^n P_n(\cos\theta)$$
$$\& \sum \frac{B_n}{r^{n+1}} P_n(\cos\theta)$$

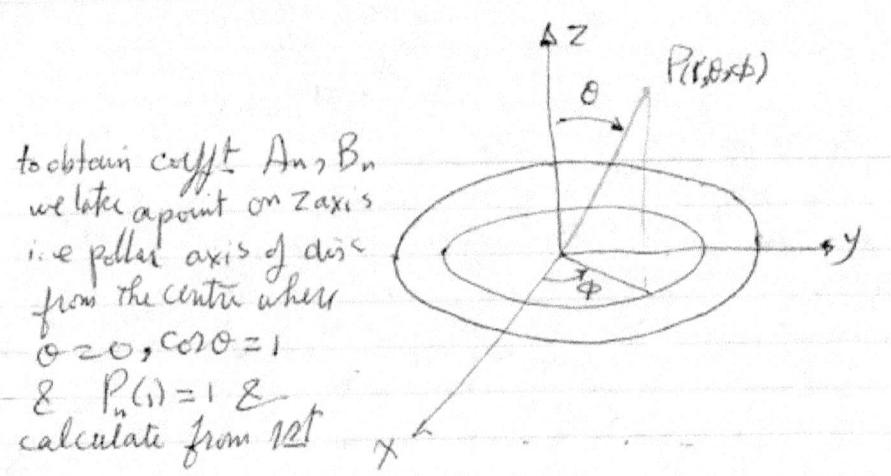

to obtain coeff.$\underline{t}$ A_n, B_n
we take a point on z axis
i.e pollar axis of disc
from the centre where
$\theta = 0$, $\cos\theta = 1$
& $P_n(1) = 1$ &
calculate from $12\underline{t}$

principales, potential at such a p$\underline{t}$ Then compare
the results with that obtained from the above expression
potential at $\underline{P}$ is

$$k\sigma \int_0^c \frac{2\pi x \, dx}{\sqrt{r^2 + x^2}}$$

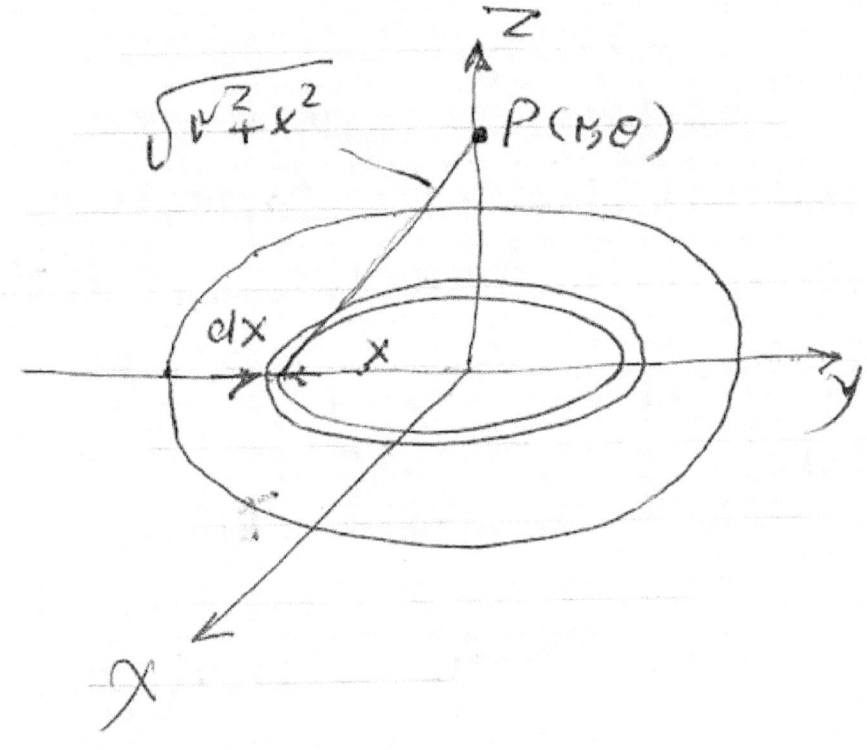

$\sqrt{r^2 + x^2}$ $P(r, \theta)$

dx x

k is the gravitational const.

$$= 2\pi k\sigma \int_0^c \frac{x\,dx}{\sqrt{r^2 + x^2}}$$

$$= \frac{2\pi k M}{(2)\,\pi c^2} \left[\sqrt{r^2 + x^2} \right]_0^c = \frac{kM}{c^2} \left[\sqrt{r^2 + c^2} - r \right]$$

$\underline{case\ I}\ r < c$ $\sqrt{r^2 + c^2} = c \left(1 + \frac{r^2}{c^2} \right)^{\frac{1}{2}} = c \left[1 + \frac{1}{2}\frac{r^2}{c^2} + \frac{1}{2} \times \frac{1}{2} \times \frac{1}{2!} \frac{r^4}{c^4} \right]$

75

$$\approx \sqrt{r^2 + c^2} = c\left(1 + \frac{1}{2}\frac{r^2}{c^2} - \frac{1}{8}\frac{r^4}{c^4} \cdots\right)$$

$$V(r,\theta) = \frac{kM}{c}\left[1 - \frac{r}{c} + \frac{1}{2}\frac{r^2}{c^2} - \frac{1}{8}\frac{r^4}{c^4} \cdots\cdots\right]$$

$$V(r,\theta) = \frac{kM}{c}\left[1 - \frac{r}{c}P_1(\cos\theta) + \frac{1}{2}\frac{r^2}{c^2}P_2(\cos\theta) - \right.$$
$$\left. \frac{1}{8}\frac{r^4}{c^4}P_4(\cos\theta)..\right]$$

<u>case II</u> $r < c$ as before similarly.

$$V(r,\theta) = \frac{kM}{c}\left[\frac{1}{2}\frac{c}{r} - \frac{1}{8}\frac{c^3}{r^3} \cdots\right]$$

$$= \frac{kM}{c}\left[\frac{1}{2}\frac{c}{r} - \frac{1}{8}\frac{c^3}{r^3}P_2(\cos\theta) \cdots\right]$$

4.6. Legendre-Fourier expansion of power function

25 <u>example</u>

expand the f^n x^3 in Lagendre-fourier

$$x^3 = a_0 P_0(x) + a_1 P_1(x) + a_2 P_2(x) + a_3 P_3(x)$$

multiply by α coefft of a_n

$$\int_{-1}^{1} x^3 P_n(x)\,dx = a_n \int P_n^2(x)\,dx$$

$$= a_n \frac{2}{2n+1}$$

$$a_n = \frac{2n+1}{2} \int_{-1}^{1} x^3 P_n(x)\, dx \qquad \text{——①}$$

$$a_0 = \frac{1}{2} \int_{-1}^{1} x^3 P_0 x\, dx \qquad \boxed{P_0(x) = 1}$$

$$= \frac{1}{2} \int_{-1}^{1} x^3 dx = \frac{1}{2} \cdot \frac{1}{4} \left[x^4 \right]_{-1}^{1} = 0 \qquad \text{——②}$$

$$a_1 = \frac{3}{2} \int_{-1}^{1} x^3 P_1(x)\, dx \qquad \boxed{P_1(x) = x}$$

$$= \frac{3}{2} \int_{-1}^{1} x^4 dx = \frac{3}{2} \left[\frac{x^5}{5} \right]_{-1}^{1} = \frac{3}{5} \qquad \text{——③}$$

$$a_2 = \frac{5}{2} \int_{-1}^{1} x^3 P_2(x)\, dx .$$

$$\therefore \boxed{(n+1) P_{n+1} - (2n+1) P_n + n P_{n-1} = 0} \quad \text{——①}$$

$\underline{put\ n = 1}$

$$\therefore\ 2 P_2 - 3 P_1 x + P_0 = 0$$

$$\boxed{P_2 = \frac{3 P_1 x - P_0}{2} = \frac{3x^2 - 1}{2}}$$

$$\therefore\ a_2 = \frac{5}{2} \int_{-1}^{1} x^3 \left(\frac{3}{2} x^2 - \frac{1}{2} \right) dx$$

77

$$a_2 = \frac{5}{2} \int_{-1}^{1} \left(\frac{3}{2}x^5 - \frac{1}{2}x^3\right) dx = 0 \quad \text{———} \quad \textcircled{A}$$

$$a_3 = \frac{7}{2} \int_{-1}^{1} x^3 P_3(x) dx$$

in $\textcircled{I}$ put $n = 2$

$$3P_3 - 5xP_2 + 2P_1 = 0$$

$$P_3 = \frac{5xP_2 - 2P_1}{3} = \frac{5x\left(\frac{3x^2-1}{2}\right) - 2x}{3}$$

$$P_3 = \frac{5}{2}x^3 - \frac{5}{6}x - \frac{2}{3}x =$$

$$\boxed{P_3 = \frac{5}{2}x^3 - \frac{9}{6}x}$$

$$a_3 = \frac{7}{2} \int_{-1}^{1} \frac{5}{2}x^6 - \frac{3}{2}x^4 dx = \frac{7}{2}\left[\frac{5}{2\times7}x^7 - \frac{3}{10}x^5\right)_{-1}^{1}$$

$$= \frac{7}{2}\left[\frac{5}{7} - \frac{3}{5}\right] = \frac{7}{2}\left(\frac{25-21}{35}\right) = \frac{2}{5} \quad \text{———} \textcircled{B}$$

$$\therefore$$

$$\boxed{x^3 = \frac{3}{5}P_1(x) + \frac{2}{5}P_3(x)}$$

5. Exercises on special functions

(1) If α is a root of the equation $J_o(x) = 0$, prove that

$$\int_0^1 J_1(\alpha x)\, dx = \frac{1}{\alpha} \quad , \quad \int_0^\alpha J_1(x)\, dx = 1$$

(2) Prove that $\displaystyle\int_0^\infty J_1(x)\, dx = 1$

(3) If $\alpha \neq 0$ is a root of the equation $J_1(x) = 0$, prove that

$$\int_0^1 x\, J_o(\alpha x)\, dx = 0$$

(4) Prove that $\displaystyle\int J_o(x)\, J_1(x)\, dx = -\frac{1}{2}\, J_o^2(x) + C$

(5) Prove that $\displaystyle\int x^2 J_o(x)\, J_1(x)\, dx = \frac{1}{2}\, x^2\, J_1^2(x) + C$

(6) Prove that $\displaystyle J_{\frac{3}{2}}(x) = \sqrt{\frac{2}{\pi x}}\left(\frac{\sin x}{x} - \cos x\right)$

(7) Prove that $y = x^n J_n(x)$ is a solution of the equation

$$xy'' + (1 - 2n)\, y' + xy = 0$$

(8) Solve the equation $\displaystyle x\, \frac{d}{dx}\left(x\, \frac{dy}{dx}\right) + k^2 x^2 y = 0$

[Ans. $y = AJ_o(kx) + BY_o(kx)$]

(9) Show that

$$x J_o''(x) + J_o'(x) + x J_o(x) = 0$$

$$x Y_o''(x) + Y_o'(x) + x Y_o(x) = 0$$

(10) Solve the equation

$$\frac{d^2\theta}{dt^2} + \frac{1}{t}\frac{d\theta}{dt} + 4t^2\theta = 0$$

$$\left[\text{Put } y = t^2, \text{ Ans. } \theta = A J_o(t^2) + BY_o(t^2) \right]$$

(11) Solve the equation

$$\frac{d^2y}{dx^2} + \left(k^2 + \frac{1}{4x^2}\right) y = 0$$

$$\left[\text{Put } y = v x^{\frac{1}{2}}, \text{ Ans. } y = x^{\frac{1}{2}} \left[A J_o(kx) + BY_o(kx) \right] \right]$$

(12) Solve the equation

$$\frac{d^2y}{dx^2} + \frac{1}{x}\frac{dy}{dx} + \frac{k^2 y}{x} = 0$$

$$\left[\text{Put } v = x^{\frac{1}{2}} \text{ Ans. } y = A J_o(2k\sqrt{x}) + BY_o(2k\sqrt{x}) \right]$$

(13) By the aid of the recurrence formulae, prove the following relations

$$J_2(x) = J_o''(x) - \frac{J_o'(x)}{x}$$

$$2 J_o''(x) = J_2(x) - J_o(x)$$

(14) Write a differential equation which has a solution
$$y = J_o'(x)$$

$$\left[\text{Ans.} \quad \frac{d^2y}{dx^2} + \frac{1}{x}\frac{dy}{dx} + \left(1 - \frac{1}{x^2}\right) y = 0 \right.$$

(15) Show that the equation

$$\frac{d^2y}{dx^2} + \frac{1}{x}\frac{dy}{dx} + \left(1 - \frac{n^2}{x^2}\right) y = 0$$

can be written in the form

$$\frac{d^2\omega}{dx^2} + \left(1 - \frac{4n^2 - 1}{4x^2}\right) \omega = 0$$

where $\omega = y\,x^{\frac{1}{2}}$

(16) Solve the equation

$$\frac{d^2y}{dx^2} + \frac{1}{x}\frac{dy}{dx} - k^2y = 0$$

[Ans. $y = A I_o (kx) + B K_o (kx)$]

(17) Solve the following equation of conduction of heat

$$\frac{d^2U}{dr^2} + \frac{1}{r}\frac{dU}{dr} - k^2U = - k^2$$

[Ans. $U = 1 + A I_o (kr) + B K_o (kr)$]

(18) In solving a problem on the distribution of the temperature U in a right circular solid cylinder we meet the
following equation

$$\frac{\partial^2U}{\partial r^2} + \frac{1}{r}\frac{\partial U}{\partial r} + \frac{\partial^2U}{\partial z^2} = 0$$

Prove that two special solutions of this equation are

$$U = J_0(k_n r) \sinh(k_n z), \quad U = J_0(k_n r) \cosh(k_n z)$$

(19) If $u_n = \int x^n J_0(x) \, dx$, prove the reduction formula

$$u_n = x^n J_1(x) + (n-1) x^{n-1} J_0(x) - (n-1)^2 u_{n-2}$$

(20) Show that the equation $\dfrac{d^2 y}{dx^2} + \dfrac{1}{x} \dfrac{dy}{dx} - y = 0$ can be

put in the form

$$\frac{d}{dx}\left(x \frac{dy}{dx}\right) = x y$$

and hence deduce that

$$\int x \, I_0(x) \, dx = x \, I_0{}'(x) + C$$

(21) Prove the following integrals

$$\int x^n J_{n-1}(\alpha x) \, dx = \frac{x^n J_n(\alpha x)}{\alpha}$$

$$\int \frac{J_{n+1}(\alpha x) \, dx}{x^n} = -\frac{1}{\alpha} \frac{J_n(\alpha x)}{x^n}$$

(22) A right circular solid cylinder of radius a is so long that its length may by regarded as infinite. It was heated to 1° C and then its surface is cooled suddenly to 0° C and kept at that temperature. Find the temperature at any instant and at any point inside the cylinder.

$$\left[\text{Ans. } U(r,t) = \sum_{n=1}^{\infty} \frac{2}{\alpha_n} \cdot \frac{1}{J_1(\alpha_n)} J_0\left(\frac{\alpha_n}{a} r\right) e^{-k \frac{\alpha_n^2}{a^2} t}\right]$$

(23) Prove the following relations

(i) $P_{2n}(-x) = P_{2n}(x)$, (ii) $P_{2n+1}(-x) = -P_{2n+1}(x)$,

(iii) $P_{2n+1}(0) = 0$

(24) The function x^3 can be expanded in the form

$$x^3 = A_0 P_0(x) + A_1 P_1(x) + A_2 P_2(x) + A_3 P_3(x)$$

Find this expansion.

$$\left[\text{Ans.} \quad x^3 = \frac{3}{5} P_1(x) + \frac{2}{5} P_3(x)\right].$$

(25) Prove that

$$\int x \log x \, J_0(x) \, dx = J_0(x) + x \log x \, J_1(x)$$

(26) Prove that

(i) $$\int_0^1 x^3 J_0(\alpha x) \, dx = \frac{\alpha^2 - 4}{\alpha^3} J_1(\alpha) + \frac{2}{\alpha^2} J_0(\alpha)$$

(ii) $$\int_0^1 x(1-x^2) J_0(\alpha x) \, dx = \frac{4}{\alpha^3} J_1(\alpha) - \frac{2}{\alpha^2} J_0(\alpha)$$

(27) Expand x^2 in a series of Bessel functions of zero order

$$\left[\text{Ans.} \quad x^2 = 2 \sum_{=1}^{\infty} \frac{\alpha_n^2 - 4}{\alpha_n^3 J_1(\alpha_n)} \cdot J_0(\alpha_n x)\right]$$

83

(28) Show that

$$J_{-\frac{3}{2}}(x) = \sqrt{\frac{2}{\pi x}} \left(-\sin x - \frac{\cos x}{x} \right)$$

(29) Show that

(i) $J_2(x) = \dfrac{2}{x} J_1(x) - J_o(x)$

(ii) $J_3(x) = \left(\dfrac{8}{x^2} - 1 \right) J_1(x) - \dfrac{4}{x} J_o(x)$

(iii) $J_4(x) = \left(\dfrac{48}{x^3} - \dfrac{8}{x} \right) J_1(x) - \left(\dfrac{24}{x^2} - 1 \right) J_o(x)$

(iv) $J_2'(x) = \dfrac{2}{x} J_o(x) + \left(\dfrac{4}{x^2} - 1 \right) J_o'(x)$

(30) A condenser microphone diaphragm is a piece of aluminium alloy foil 2.54×10^{-3} cm. thick stretched over a metal framework 4 cm. diameter. What must be the total radial tension if the fundamental frequency in vacuo is 5000 $\sim$ and the density of the foil is 2.7 gm./cm.^{3} ?

[Ans. 5.9×10^7 dynes]

By the aid of Beta and Gamma functions prove the following

(31) $\displaystyle\int_0^{\frac{\pi}{2}} \sin^5\theta \cos^3\theta \, d\theta = \dfrac{1}{24}$

84

(32) $\displaystyle\int_0^{\frac{\pi}{2}} \sin^6 \theta \cos^2 \theta \ d\theta = \frac{5\pi}{256}$

(33) $\displaystyle\int_0^{\frac{\pi}{2}} \cos^4 \theta \sin^5 \theta \ d\theta = \frac{8}{315}$

(34) $\displaystyle\int_0^{\frac{\pi}{2}} \sin^2 \theta \cos^4 \theta \ d\theta = \frac{\pi}{32}$

(35) $\displaystyle\int_0^{\frac{\pi}{2}} \cos^6 \theta \ d\theta = \frac{5\pi}{32}$

(36) $\displaystyle\int_0^{\frac{\pi}{2}} \sin^5 \theta \ d\theta = \frac{8}{15}$

(37) $\displaystyle\int_0^{\frac{\pi}{2}} \sin^3 \theta \cos^4 \theta \ d\theta = \frac{2}{35}$

(38)   If m and n are positive integers, prove that

$$\beta(m,n) = \frac{(m-1)! \ (n-1)!}{(m+n-1)!}$$

CHAPTER 8

FOURIER TRANSFORMS

1. Fourier series and harmonic analysis

Fourier Series and Harmonic Analysis

Periodic fns

$$\sin x, \cos x \qquad 2\pi \qquad \sin(x + 2\pi) = \sin x$$
$$\tan x, \cot x \qquad \pi \qquad \tan(x + \pi) = \tan x$$
$$\sin(nx), \cos(nx) \qquad \frac{2\pi}{n}$$

A given fn is said to be a periodic of period l if
$$f(x+l) = f(x).$$

i.e. the additional of l to x leaves the fn unchanged
it is clear that if l is a period then any *integral* multiply by l
is also a period

The graph of a periodic fn repeats itself for every period
examples of periodic fns are.

1. $\sin x, \cos x$, period 2π, $\sin(x + 2\pi) = \sin x$
2. $\tan x, \cot(x)$, period π, $\tan(x + \pi) = \tan x$.
3. $\sin nx, \cos nx$, N, $\frac{2\pi}{n}$, for $\sin n(x + \frac{2\pi}{n}) = \sin(nx + 2\pi) = \sin nx.$
4. the current & voltage in an alternating current circuit.
5. the displacement velocity of the position of the Heat engine

2. Fourier theorem

86

Fourier theorem

A period f^n $y = f(x)$ of period 2π can be expanded in the form $y = f(x) = a_0 + a_1 \cos x + a_2 \cos 2x + \cdots + a_n \cos^n$

$+ b_1 \sin x + b_2 \sin 2x + \cdots + b_n \sin nx.$

($*$ A periodic f^n $y = f(x)$ of period 2π can be expanded in the form

$y = f(x) = a_0 + a_1 \cos x + a_2 \cos 2x + \cdots + a_n \cos nx.$

$+ b_1 \sin x + b_2 \sin 2x + \cdots + b_n \sin nx.$))

conditions

1- $f(x)$ is bounded that is never _infinite_ in one period.

2. bet^n any two values of x in one period the number maximum or minimum is finite & the number of discontinuity is also finite

This ats conditions are usually satisfied in almost all practical properties

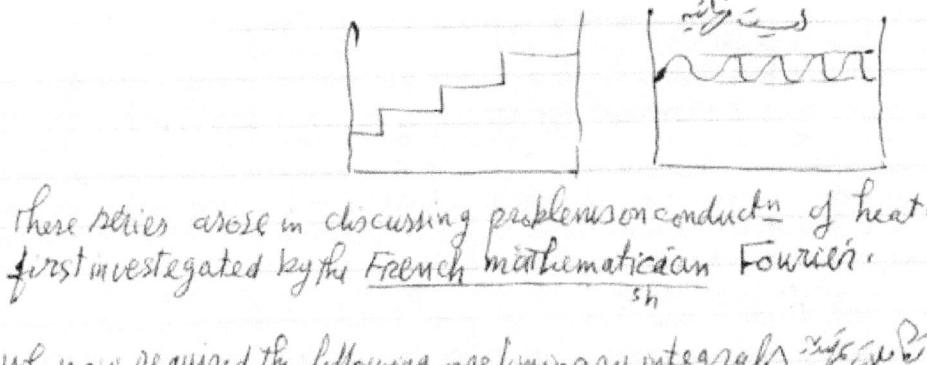

These series arose in discussing problems on conduct^n of heat. first investigated by the French mathematician Fourier.

we now require the following preliminary integrals

3. Preliminary integrals used in Fourier transforms

Preliminary Integrals

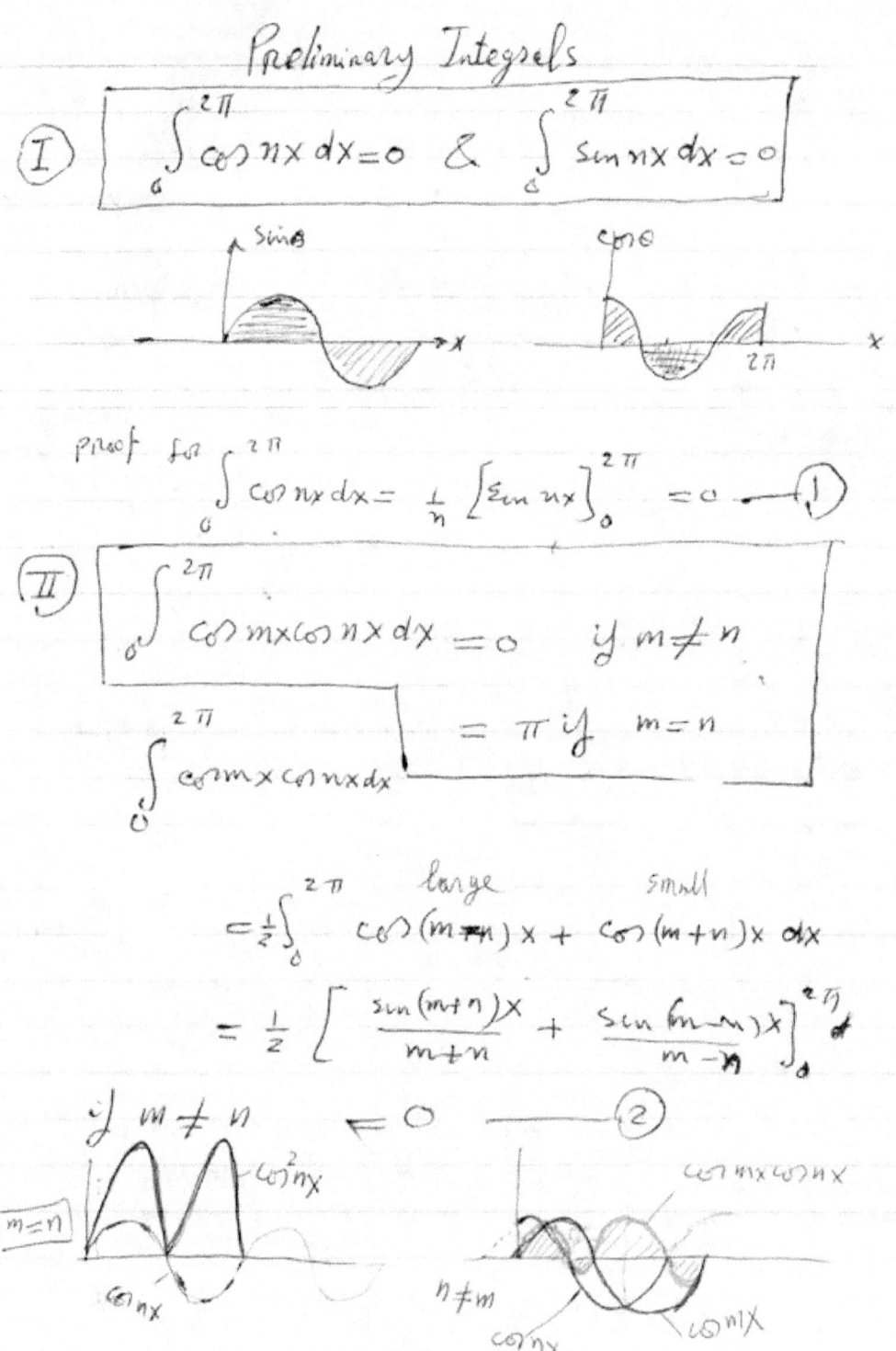

(I)
$$\int_0^{2\pi} \cos nx \, dx = 0 \quad \& \quad \int_0^{2\pi} \sin nx \, dx = 0$$

proof for $\int_0^{2\pi} \cos nx \, dx = \frac{1}{n} \left[\sin nx \right]_0^{2\pi} = 0 \quad \text{———(1)}$

(II)
$$\int_0^{2\pi} \cos mx \cos nx \, dx = 0 \quad \text{if } m \neq n$$
$$= \pi \text{ if } m = n$$

$\int_0^{2\pi} \cos mx \cos nx \, dx$

$$= \frac{1}{2} \int_0^{2\pi} \overset{large}{\cos(m-n)x} + \overset{small}{\cos(m+n)x} \, dx$$

$$= \frac{1}{2} \left[\frac{\sin(m+n)x}{m+n} + \frac{\sin(m-n)x}{m-n} \right]_0^{2\pi}$$

if $m \neq n$ $\quad = 0 \quad$ ———(2)

$m = n$

if $m=n$ $\therefore$ $\displaystyle\int_0^{2\pi} \cos^2 nx\, dx = \frac{1}{2}\int_0^{2\pi}(1+\cos 2nx)\, dx$

$$= \frac{1}{2}\left[x + \frac{\sin 2nx}{2n}\right]_0^{2\pi}$$

$$= \pi \underline{\qquad\qquad} ③$$

Simillarity

$$\boxed{\int_0^{2\pi} \sin mx\, \sin nx\, dx = 0 \quad n\neq m \\ \qquad\qquad\qquad = \pi \quad n = m}$$

Ⅶ $\quad\boxed{\displaystyle\int_0^{2\pi} \sin mx\, \cos nx\, dx = 0 \quad weither\ m=n \\ \qquad\qquad\qquad\qquad\qquad or\ m\neq n}$

$$\frac{1}{n}\int_0^{2\pi} \sin mx\, d\sin nx$$

if $= m = n$ $\qquad = \frac{1}{n}\left[\frac{\sin^2 mx}{2}\right]_0^{2\pi} = 0 \quad ④$

ii) $m\neq n$
$\therefore$ $\displaystyle\int_0^{2\pi} \sin mx\, \cos nx\, dx = \frac{1}{2}\int_0^{2\pi}\sin(m+n)x + \sin(m-n)x$

$$= \frac{1}{2}\left[-\frac{\cos(m+n)x}{m+n} - \frac{\cos(m-n)x}{m-n}\right]_0^{2\pi}$$

$$= \frac{1}{2}\left[-\frac{1}{m+n} - \frac{1}{m-n} + \frac{1}{m+n} + \frac{1}{m-n}\right] = 0 \quad ⑤$$

4. Determination of the coefficients of the Fourier expansion

Determination of the coeff.^{ts} of the expansion.

$$y = f(x) = a_0 + a_1 \cos x + a_2 \cos 2x + \cdots \quad + a_n \cos nx$$
$$+ b_1 \sin x + b_2 \sin 2x + \cdots + b_n \sin nx.$$

evaluation of a_0:

integrate both sides w.r.t x bet zero & 2π.

$$\int_0^{2\pi} f(x)\,dx = a_0 \int_0^{2\pi} dx = a_0\, 2\pi \cdot \quad ① $$

all other integrals vanish.

$$\boxed{a_0 = \frac{1}{2\pi} \int_0^{2\pi} f(x)\,dx.}$$

∴ a_0 is the mean value of $f(x)$ in the interval $0 \leq x < 2\pi$

i.e. a_0 is the height of the rectangular whose base is the period & whose area is equal to the area under the curve.

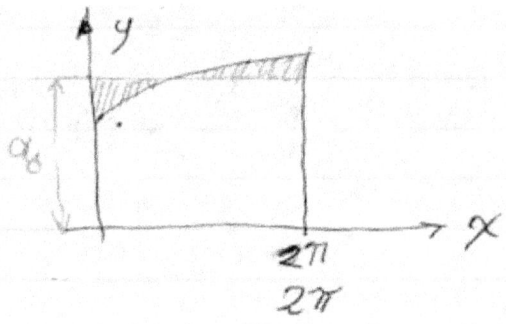

<u>Multiply</u> both sides by the coeff^t of a_n which is $\cos nx$ & integrate bet^n zero & 2π.

$$\int_0^{2\pi} f(x) \cos nx \, dx = a_n \int_0^{2\pi} \cos^2 nx \, dx = \pi a_n \quad —②$$

$$a_n = \frac{1}{\pi} \int_0^{2\pi} f(x) \cos nx \, dx$$

Similarly

$$b_n = \frac{1}{\pi} \int_0^{2\pi} f(x) \sin nx \, dx$$

<u>example</u>

Analyse the following fn.

$$f(x) = a \qquad 0 < x < \pi$$
$$= 0 \qquad \pi < x < 2\pi$$

Sol^n

$$a_0 = \frac{a \times \pi}{2\pi} = \frac{a}{2} \quad —①$$

$$a_n = \frac{1}{\pi} \int_0^{2\pi} f(x) \cos nx \, dx \quad —②$$

evaluation of a_n

Multiply both sides by the coeff^t of a_n which is $\cos nx$ & integrate bet^n zero & 2π.

$$\int_0^{2\pi} f(x) \cos nx \, dx = a_n \int_0^{2\pi} \cos^2 nx \, dx = \pi\, a_n \quad —— (2)$$

$$\boxed{\begin{array}{l} a_n = \dfrac{1}{\pi} \displaystyle\int_0^{2\pi} f(x) \cos nx \, dx \\[4mm] b_n = \dfrac{1}{\pi} \displaystyle\int_0^{2\pi} f(x) \sin nx \, dx \end{array}}$$

Similarly

5. Examples of Fourier transformations

example

Analyse the following fn.
$$f(x) = a \qquad 0 < x < \pi$$
$$= 0 \qquad \pi < x < 2\pi$$

Sol^n

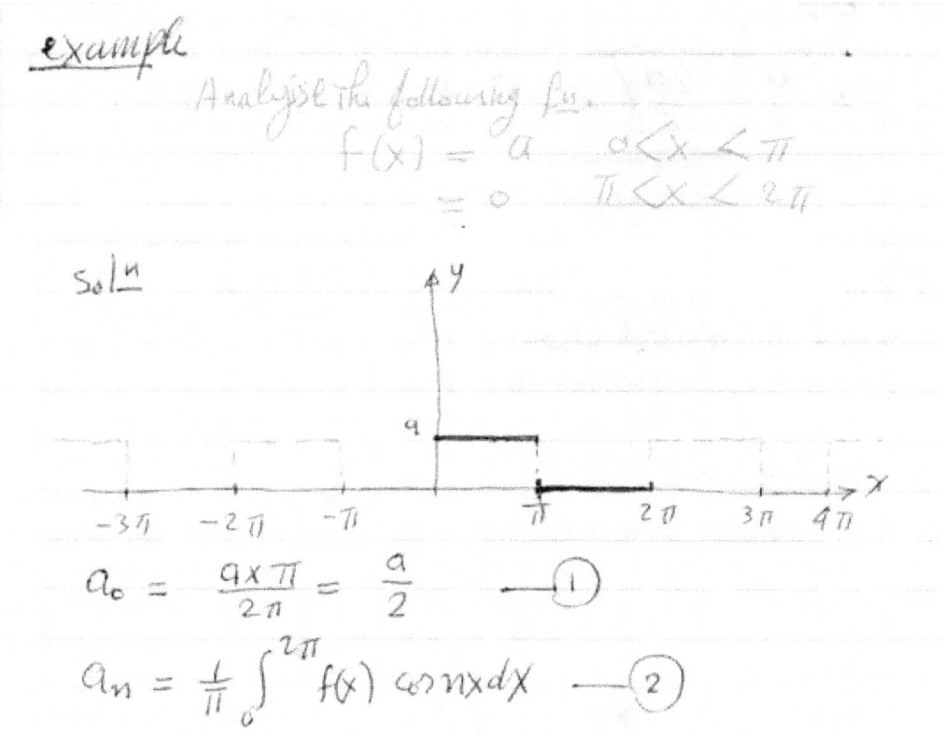

$$a_0 = \frac{a \times \pi}{2\pi} = \frac{a}{2} \qquad \text{①}$$

$$a_n = \frac{1}{\pi} \int_0^{2\pi} f(x) \cos nx \, dx \qquad \text{②}$$

$$\therefore a_n = \frac{1}{\pi} \int_0^{\pi} a \cos nx \, dx = \frac{a}{n\pi} \left[\sin nx \right]_0^{\pi}$$

$$a_n = \frac{a}{n\pi} \{ 0 \} = 0 \qquad \text{②'}$$

$$b_n = \frac{1}{\pi} \int_0^{2\pi} f(x) \sin nx \, dx = \frac{1}{\pi} \int_0^{\pi} a \sin nx \, dx$$

$$b_n = \frac{-a}{n\pi} \left[\cos nx \right]_0^{\pi} = \frac{-a}{n\pi} \left[\cos n\pi - 1 \right] \qquad \text{③}$$

93

$$b_n = \frac{2a}{n\pi} \qquad (n)\,odd \qquad \text{ش}$$

$$b_n = 0 \qquad (n)\,even. \qquad \text{ش}$$

$$\therefore \boxed{y = \frac{a}{2} + \frac{2a}{\pi}\left[\sin x + \frac{1}{3}\sin 3x + \frac{1}{5}\sin 5x + \cdots\right]}$$

We shall now consider the modification of the above formulii to suite the special cases.

6. Fourier expansions in cosines only

I. Expansions in cosines only.

i.e $\quad y = a_0 + a_1 \cos x + a_2 \cos 2x + \cdots + a_n \cos nx.$

This occurs when $\boxed{f(-x) = f(x)}$ that is the analysis

curve is symmetrical about the $\underline{y-axis}$

$$a_n = \frac{1}{\pi} \int_0^{2\pi} f(x) \cos nx \; dx.$$

94

$$= \frac{1}{\pi} \int_{-\pi}^{\pi} f(x) \cos nx \, dx$$

even x even = even.

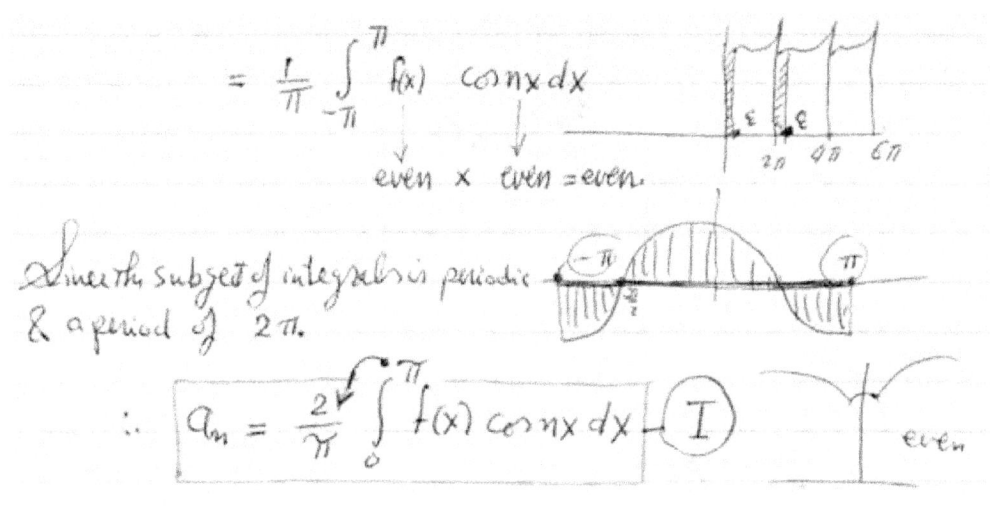

Since the subject of integrals is periodic & a period of 2π.

$$\therefore \quad a_n = \frac{2}{\pi} \int_{0}^{\pi} f(x) \cos nx \, dx \quad \boxed{\text{I}}$$

even

Since the subject of integrals is an even f^n of $f(x)$ & hence symmetrical about the y-axis in this way the range of integration is halved being reduced from (2π) to (π) & in most cases we $\dot{w}\dot{e}$ need not divide the range of integration

example express as a fourier series.

$$F(X) = X \qquad 0 < x < \pi$$
$$= -X \qquad -\pi < X < 0$$

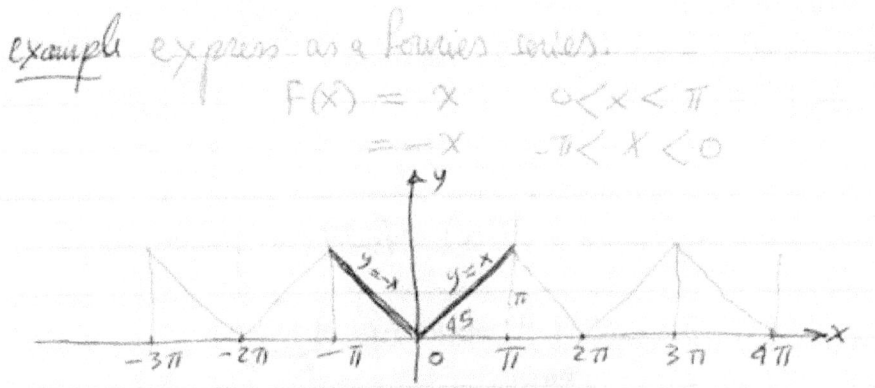

here the graph of the f^n is symmetrical about the y-axis & hence we know that the expansion contains cosines only.

soln

$$a_0 = \frac{2\pi \times \pi}{2 \times 2\pi} = \frac{\pi}{2} \quad —①$$

$$a_n = \frac{2}{\pi} \int_0^{\pi} f(x) \cos nx \, dx \quad —②$$

$$= \frac{2}{\pi} \int_0^{\pi} x \cos nx \, dx$$

by partial

$$a_n = \frac{2}{n\pi} \int_0^{\pi} x \, d\sin nx = \frac{2}{n\pi} \left[x \sin nx + \frac{\cos nx}{n} \right]_0^{\pi}$$

$$a_n = \frac{2}{n\pi} \left[\frac{\cos nx}{n} \right]_0^{\pi}$$

(n) even $\quad a_n = \frac{2}{n^2\pi} \left[(1-1) \right] = 0 \quad \Big]$

(n) odd $\quad a_n = \frac{-4}{n^2 \pi} \qquad ③$

$$y = f(x) = \frac{\pi}{2} - \frac{4}{\pi} \left[\frac{\cos x}{1^2} + \frac{\cos 3x}{3^2} + \frac{\cos 5x}{5^2} + \cdots \right]$$

cor: ..ﺳﻲ

For the graph $y=0, x=0$ this should be the case from the above expansion.

$$0 = \frac{\pi}{2} - \frac{4}{\pi} \left[\frac{1}{1^2} + \frac{1}{3^3} + \frac{1}{5^2} + \frac{1}{7^3} + \cdots \right]$$

$$\therefore \boxed{\frac{1}{1^2} + \frac{1}{3^2} + \frac{1}{5^2} + \frac{1}{7^2} + \dots \infty = \frac{\pi^2}{8}}$$

(II) <u>Expansions in sines only</u>

$$y = f(x) = b_1 \sin x + b_2 \sin 2x + \dots + b_n \sin nx.$$

7. Fourier expansions in sines only

Here

$$\boxed{f(-x) = -f(x)} \quad i.e \quad f(x) \text{ is an } \underline{odd} \; f_n$$

of x & the analysis curve is symmetrical about the <u>origin</u>. Similar to the previous case we have.

$$\boxed{b_n = \frac{2}{\pi} \int_0^{\pi} f(x) \sin nx \, dx}$$

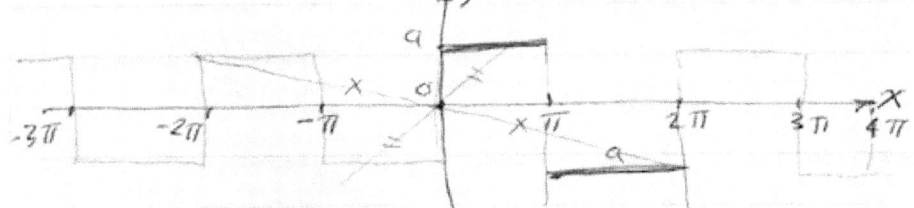

Example expand as a fourier series

$$y = a \qquad o < x < \pi$$
$$= -a \qquad \pi < x < 2\pi$$

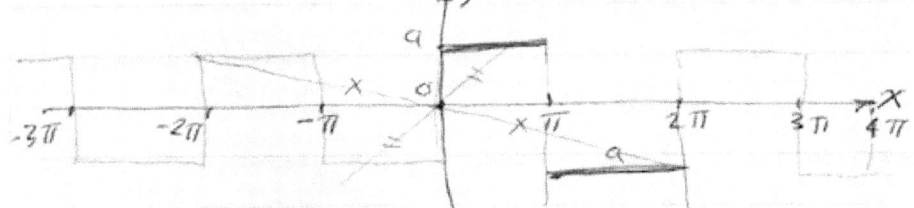

The meaning of symmetry about الدوال بعضها بعض الدوال بعض الدوال , أي أن the origin here the expansion is therefore in _sines only_

$$b_n = \frac{2}{\pi} \int_0^{\pi} f(x) \sin nx \, dx$$

$$= \frac{2}{\pi} \int_0^{\pi} a \sin nx \, dx$$

$$b_n = -\frac{2a}{n\pi} \Big[\cos nx \Big]_0^{\pi} = +\frac{4a}{n\pi} \quad (n) \text{ odd}$$

$$= o \quad (n) \text{ even}$$

$$y = +\frac{4a}{\pi} \left[\frac{\sin x}{1} + \frac{\sin 3x}{3} + \frac{\sin 5x}{5} + \ldots \ldots n \right]$$

what is the meaning expansion?
it gives a general eq^n for a straight horizontal lines which is impossible in the analytic geometry وذلك لأن

<u>Note</u>: At a point of <u>discontinuity</u> عند نقطة the expansion gives the mean value of the f^n <u>Just before</u> & <u>Just after</u> the point of discontinuity.

Now we try to explain the meaning of this result.

1st approximation $y = \frac{4a}{\pi} \sin x$ ————— ①

2nd approximation $y = \frac{4a}{\pi} \left[\sin x + \frac{1}{3} \sin 3x \right]$ —— ②

3rd $\sim$ $y = \frac{4a}{\pi} \left[\sin x + \frac{1}{3} \sin 3x + \frac{1}{5} \sin 5x \right]$ ③

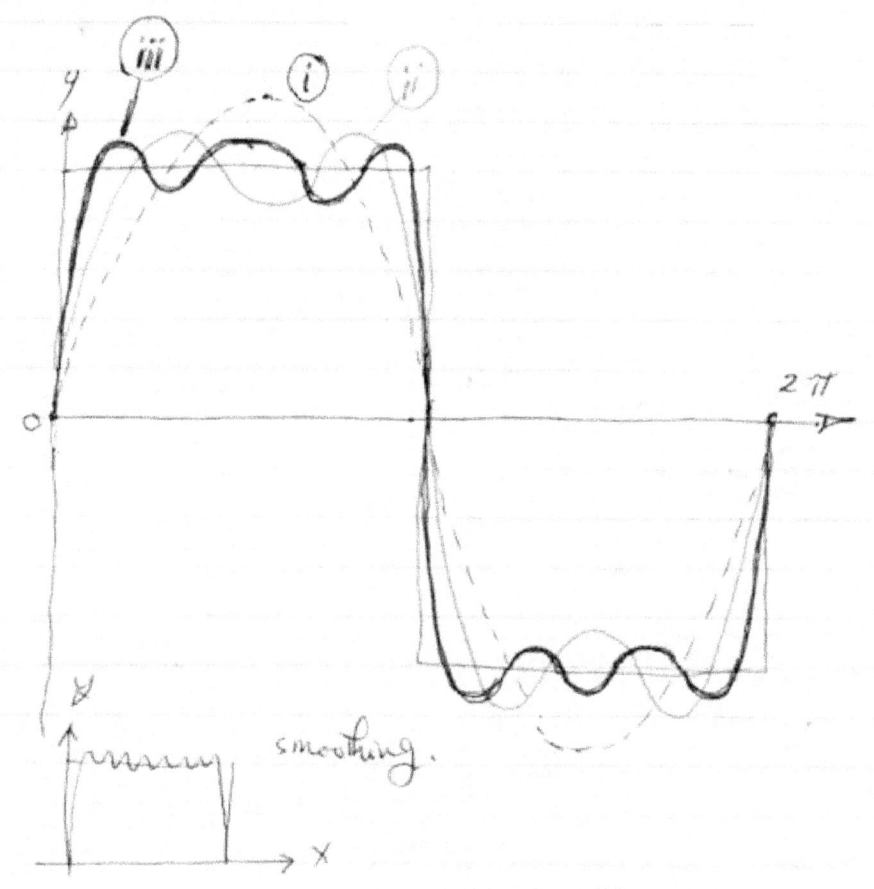

smoothing.

8. Fourier expansions in even harmonics

22\9\1972

III Expansions in Even Harmonics

$$y = f(x) = a_0 + a_2 \cos 2x + a_4 \cos 4x + \cdots + a_{2n} \cos 2nx$$
$$+ \ b_2 \sin 2x + b_4 \sin 4x + \cdots + b_{2n} \sin 2nx$$

This expansion occurs when $\boxed{f(x + \pi) = f(x)}$ i.e the curve repeated itself every π.

$\dfrac{\tan x}{\cot x}$

$$a_{n(even)} = \frac{1}{\pi} \int_0^{2\pi} f(x) \cos nx \, dx$$

$$a_{n(even)} = \frac{1}{\pi} \int_0^{\pi} f(x) \cos nx \, dx + \frac{1}{\pi} \int_\pi^{2\pi} f(x) \cos nx \, dx$$

Consider the integral $\frac{1}{\pi} \int_\pi^{2\pi} f(x) \cos nx \, dx$. but $x = \pi + z$
$$dx = dz$$

δ integral reduces to.

$$\frac{1}{\pi} \int_\pi^{2\pi} f(x) \cos nx \, dx = \frac{1}{\pi} \int_0^{\pi} f(z + \pi) \cos n(z+\pi) \, dz$$

$x = \pi, z = 0$
$x = 2\pi, z = \pi$

$$= \frac{1}{\pi} \int_0^{\pi} f(z) \cos(nz) \cdot dz.$$

Since $f(x)$ is periodic & of period π & n is even. Hence

$$a_{n(even)} = \frac{2}{\pi} \int_0^\pi f(x) \cos nx\, dx$$

$$b_{n(even)} = \frac{2}{\pi} \int_0^\pi f(x) \sin nx\, dx$$

__example__   express as fourier series

$$F(x) = x \qquad 0 < x < \pi$$
$$= x - \pi \qquad \pi < x < 2\pi$$

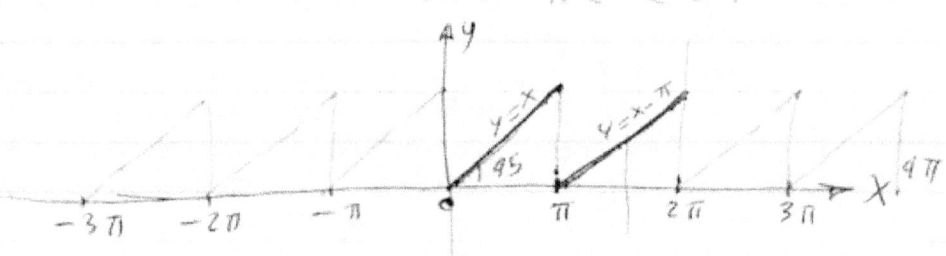

here the curve repeats its self every π & hence the expansion is in __even harmonics__ ـ أي الزوجية since __odd__ Harmonics vanishes

because
$$a_n = \frac{2}{\pi} \int_0^\pi f(x) \cos nx\, dx$$
$$= \frac{2}{\pi} \int_0^\pi x \cos nx\, dx$$
$$= \frac{2}{n\pi} \int_0^\pi x\, d(\sin nx)$$

$$= \frac{2}{n\pi} \left[x \sin nx + \frac{\cos nx}{n} \right]_0^{\pi}$$

n odd $a_n = \frac{2}{n^2 \pi} [-1 -1] = -\frac{4}{n^2 \pi}$

n even $a_n = \frac{2}{n^2 \pi} [1 - 1] = 0$

hence

odd × even = even ≠ 0

even × even = even = 0

sol^n

$$a_0 = \frac{\pi \times \pi}{2 \pi} = \frac{\pi}{2} \qquad \text{—①}$$

$$a_{n(even)} = \frac{2}{\pi} \int_0^{\pi} f(x) \cos(nx) dx$$

$$= \frac{2}{\pi} \int_0^{\pi} x \cos nx \, dx$$

$$= \frac{2}{n\pi} \int_0^{\pi} x \, d\sin nx = \frac{2}{n\pi} \left[x \sin nx + \frac{\cos nx}{n} \right]_0^{\pi}$$

$$a_{n\,(even)} = \frac{2}{n^2 \pi} \left[\cos n\pi - 1 \right] = 0 \quad \text{since (n is even)} \qquad \text{②}$$

$$b_{n(even)} = \frac{2}{\pi} \int_0^{\pi} f(x) \sin nx \, dx = -\frac{2}{\pi} \int_0^{\pi} x \, d\cos nx \, dx$$

$$= -\frac{2}{n\pi}\left[x\cos nx - \frac{\sin nx}{n}\right]_0^{\pi}$$

$$= -\frac{2}{n\pi}\left[\pi\cos n\pi\right] = -\frac{2}{n}$$

$$y = \frac{\pi}{2} - 2\left[\frac{\sin 2x}{2} + \frac{\sin 4x}{2} + \frac{\sin 6x}{6} + \cdots\right]$$

التي يطلبها السؤال

9. Fourier expansions in odd harmonics

IV Expansion in Odd Harmonics

$$y = f(x) = a_1\cos x + a_3 \cos 3x + \cdots$$
$$b_1 \sin x + b_3 \sin 3x + \cdots$$

This expansion occurs when $\boxed{f(x+\pi) = -f(x)}$
i.e. the curve repeats itself
every π with an opposite sign كما هو موجود & as the previous case.

$$a_n (odd) = \frac{2}{\pi}\int_0^{\pi} f(x)\cos nx\,dx$$

$$b_n(odd) = \frac{2}{\pi}\int_0^{\pi} f(x)\sin nx\,dx$$

For example in the previous example. $y = a \quad 0 < x < \pi$
in this case the curve repeats its self $\qquad = -a \quad \pi < x < 0$
every π with an opposite sign.

103

i.e the expansion is odd Harmonic also. we have symmetry about the origin. Therefore the expansion is sines only.
combining both we see that. The expansion is in odd sines

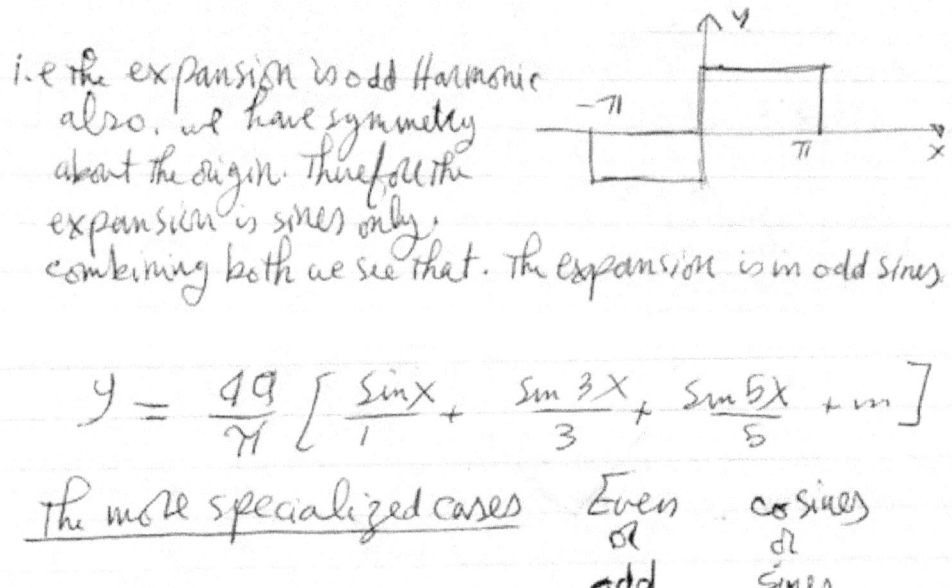

$$y = \frac{49}{\pi}\left[\frac{\sin x}{1} + \frac{\sin 3x}{3} + \frac{\sin 5x}{5} + \cdots\right]$$

The more specialized cases   Even   cosines
                               or   or
                              odd   sines

10. Summary of common Fourier transforms

i) Even cosines

cosine	$f(-x) = f(x)$
even Harmonic	$f(x + \pi) = f(x)$

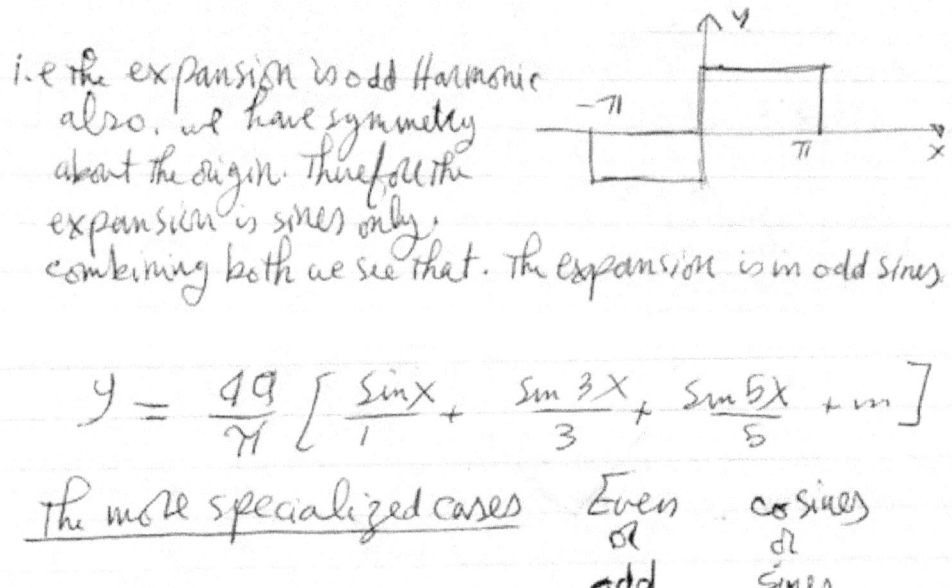

$$y = a_0 + a_2 \cos 2x + a_4 \cos 4x + \cdots$$

ii) Even sines

sines	$f(-x) = - f(x)$
even Harmonics	$f(x + \pi) = f(x)$

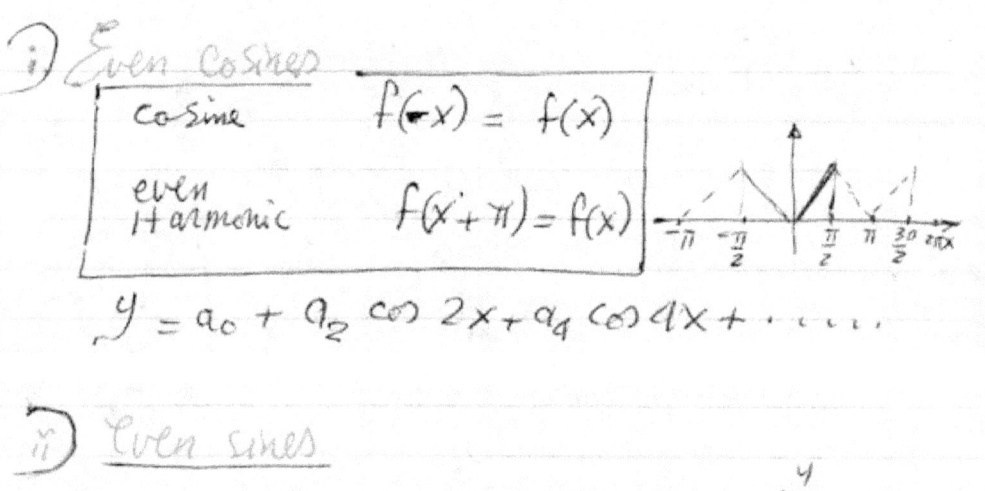

iii) <u>Odd cosines</u>

Cosines $\quad f(-x) = f(x)$
odd Harmonics $\quad f(x+\pi) = -f(x)$

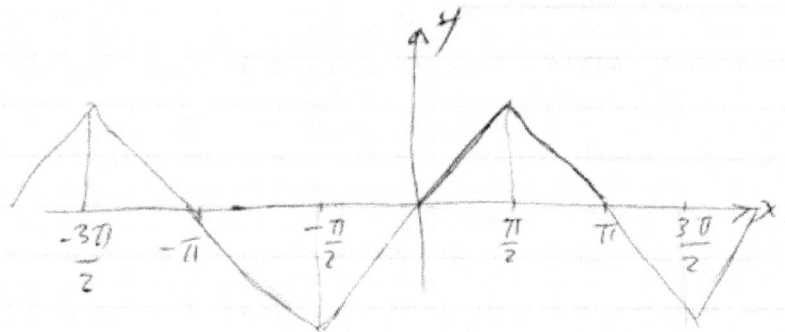

iv) <u>Odd sines</u>

sines $\quad f(-x) = -f(x)$
odd Harmonics $\quad f(x+\pi) = -f(x)$

the simplified formulii for the above four cases.
consider for example the case of even cosines.

cosine $f(-x) = f(x)$

even Harm. $f(x + \pi) = f(x)$

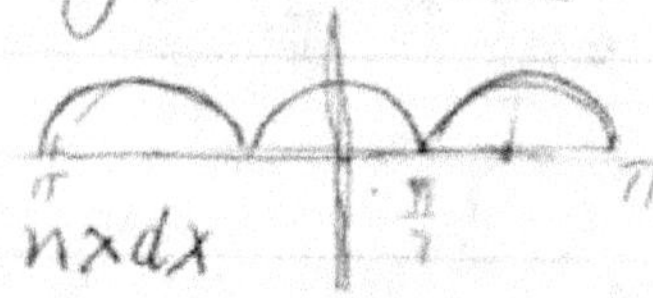

a_n (even) $= \dfrac{2}{\pi} \displaystyle\int_0^{\pi} f(x) \cos nx \, dx$

$$a_n (\text{even}) = \dfrac{2}{\pi} \int_{-\frac{\pi}{2}}^{\frac{\pi}{2}} f(x) \cos nx \, dx$$

even × even = even

Since $f(x)$ is periodic & of period π & n is even

$$\boxed{a_n = \dfrac{4}{\pi} \int_0^{\frac{\pi}{2}} f(x) \cos nx \, dx}$$

where similar formulii for the remaining 3-cases.

The interval of integration is now reduced to one quarter of a period.

example the sketch shown the teeth of a saw
$y = x$ where $0 < x < \dfrac{\pi}{2}$ find its fourier expansion.

106

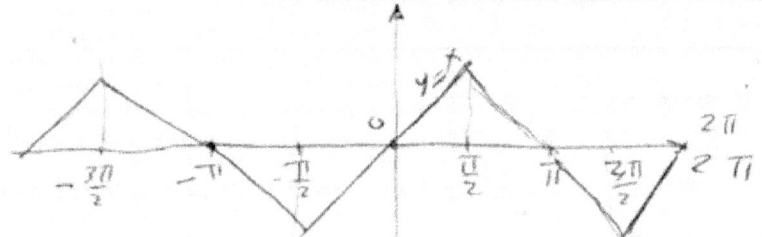

Here we have symmetry about the origin the expansion is thus
in <u>sines</u> . ———— ①

Again the self repeat itself every π with an <u>opposite</u>
<u>sign</u>
∴ expansion is an <u>odd Harmonics</u> ——→ ②
combining both the expansions in odd sines.

$$y = b_1 \sin x + b_3 \sin 3x + b_5 \sin 5x + \dots$$

$$b_{n(odd)} = \frac{4}{\pi} \int_0^{\frac{\pi}{2}} x \sin nx \, dx = -\frac{4}{n\pi} \int_0^{\frac{\pi}{2}} x \, d\cos nx$$

$$= -\frac{4}{n\pi} \left[x \cos nx + \frac{\sin nx}{n} \right]_0^{\frac{\pi}{2}}$$

$$b_n(odd) = \frac{4}{n^2 \pi} \left[\sin nx \right]_0^{\frac{\pi}{2}} = \frac{4}{n^2 \pi} \cdot \sin \frac{n\pi}{2}$$

$$= \frac{4}{n^2 \pi} \quad or = -\frac{4}{n^2 \pi}$$

$$b_{n(odd)} = \frac{4}{n^2 \pi} \qquad 1, 5, 9, \cdots$$

$$= \frac{-4}{n^2 \pi} \qquad 3, 7, 11, \cdots$$

$$y = \frac{4}{\pi}\left[\frac{\sin x}{1^2} - \frac{\sin 3x}{3^2} + \frac{\sin 5x}{5^2} - \cdots \right]$$

<u>Notice</u> if we put $x = \frac{\pi}{2}$

$$y = \frac{4}{\pi}\left\{ \frac{1}{1^2} + \frac{1}{3^2} + \frac{1}{5^2} + \cdots \right]$$

$$= \frac{4}{\pi}\left[\frac{\pi^2}{8} \right] = \frac{\pi}{2}$$

which is the same value as the graph

<u>change of period</u>:

In all our previous work ~~with the period was 2π~~ ~~expansion now takes the form~~ suppose the period is
λ. Suppose the period is λ the expansion now takes
the form

$$y = a_0 + a_1 \cos \omega x + a_2 \cos 2\omega x + \cdots$$
$$+ b_1 \sin \omega x + b_2 \sin 2\omega x + \cdots$$

∴ the period
$$\lambda = \frac{2\pi}{\omega} \quad \& \quad \omega = \frac{2\pi}{\lambda}$$

and the expansion takes the form

$$y = f(x) = a_0 + a_1 \cos \frac{2\pi}{\lambda} x + a_2 \cos \frac{4\pi x}{\lambda} + \cdots$$
$$a_n \cos \frac{2n\pi x}{\lambda}$$
$$+ b_1 \sin \frac{2\pi x}{\lambda} + b_2 \sin \frac{4\pi x}{\lambda} + \cdots + b_n \sin \frac{2\pi n x}{\lambda}$$

i.e. we change x into $\frac{2\pi}{\lambda} x$ to obtain a_n.

we multiply both sides by $\cos \frac{2n\pi}{\lambda} x$ & integrate
bet^n o & λ

$$\int_0^\lambda f(x) \cos \frac{2n\pi}{\lambda} x \, dx = a_n \int^\lambda \cos^2 \frac{2n\pi x}{\lambda} \, dx$$

$$= \frac{a_n}{2} \int_0^\lambda \left(1 + \cos \frac{4n\pi x}{\lambda} \right) dx$$

$$= \frac{a_n}{2} \left[x + \frac{\sin \frac{4n\pi x}{\lambda}}{\frac{4n\pi}{\lambda}} \right]_0^\lambda \quad \to 0$$

$$= \frac{a_n \lambda}{2}$$

$$\therefore \boxed{a_n = \frac{2}{\lambda} \int_0^\lambda f(x) \cos \frac{2n\pi x}{\lambda} \, dx}$$

To check put $x = 2\pi$ similarly b

$$b_n = \frac{2}{\lambda} \int_0^\lambda f(x) \sin \frac{2n\pi}{\lambda} x \cdot dx$$

11. Practical Fourier Analysis

Practical Fourier Analysis 27\4\92

Preliminary theorems :- بادئ نظریات

11.1. Preliminary theorems

(I)
$$\cos\alpha + \cos(\alpha+\beta) + \cos(\alpha+2\beta) + \ldots + \cos(\alpha + \overline{n-1}\beta)$$
$$= \frac{\cos\left(\alpha + \frac{n-1}{2}\beta\right) \sin \frac{n\beta}{2}}{\sin \frac{\beta}{2}}$$

Proof: let $S = \cos\alpha + \cos(\alpha+\beta) + \cos(\alpha+2\beta) + \ldots + \cos(\alpha+\overline{n-1}\beta)$

multiply by $2 \sin \frac{\beta}{2}$

$$2S \sin \frac{\beta}{2} = 2 \cos\alpha \sin\frac{\beta}{2} + 2\cos(\alpha+\beta)\sin\frac{\beta}{2} + \ldots$$
$$+ 2\cos(\alpha+2\beta)\sin\frac{\beta}{2} + \ldots + 2\cos(\alpha+\overline{n-1}\beta) \sin\frac{\beta}{2}$$
$$= \sin\left(\alpha+\frac{\beta}{2}\right) - \sin\left(\alpha-\frac{\beta}{2}\right) +$$
$$+ \sin\left(\alpha+\frac{3\beta}{2}\right) - \sin\left(\alpha+\frac{\beta}{2}\right) + \ldots$$
$$+ \sin\left(\alpha+\frac{5}{2}\beta\right) - \sin\left(\alpha+\frac{3}{2}\beta\right) + \ldots$$
$$\sin\left[\alpha+(n-\tfrac{1}{2})\beta\right] - \sin\left(\alpha+(n-\tfrac{3}{2})\beta\right)$$

110

$$= \sin\left[\alpha + \left(n - \tfrac{1}{2}\right)\beta\right] - \sin\left(\alpha - \tfrac{\beta}{2}\right)$$

$$= 2 \cos \frac{2\alpha + (n-1)\beta}{2} \sin \frac{n}{2}\beta$$

$$= 2 \cos\left(\alpha + \tfrac{n-1}{2}\beta\right) \sin \tfrac{n}{2}\beta$$

$$\therefore \quad S = \frac{\cos\left(\alpha + \tfrac{n-1}{2}\beta\right) \sin \tfrac{n}{2}\beta}{\sin \tfrac{\beta}{2}}$$

$$\therefore \cos\alpha + \cos(\alpha+\beta) + \cos(\alpha+2\beta) + \cdots + \cos(\alpha + \overline{n-1}\,\beta)$$

$$= \frac{\cos\left(\alpha + \tfrac{n-1}{2}\beta\right) \sin \tfrac{n}{2}\beta}{\sin \tfrac{\beta}{2}} \quad \text{\textcircled{1}}$$

add $2\,\tfrac{\pi}{2}$ to both sides.

$$\cos\left(\alpha + \tfrac{\pi}{2}\right) + \cos\left(\alpha + \tfrac{\pi}{2} + 2\beta\right) + \cdots + \cos\left(\alpha + \tfrac{\pi}{2} + \overline{n-1}\,\beta\right)$$

$$= \frac{\cos\left(\alpha + \tfrac{\pi}{2} + \tfrac{n-1}{2}\beta\right) \sin \tfrac{n}{2}\beta}{\sin \tfrac{\beta}{2}}$$

$$\therefore \boxed{\begin{array}{c} \sin\alpha + \sin(\alpha+\beta) + \sin(\alpha+2\beta) + \cdots + \sin(\alpha + \overline{n-1}\,\beta) \\[2mm] = \dfrac{\sin\left(\alpha + \tfrac{n-1}{2}\beta\right) \sin \tfrac{n}{2}\beta}{\sin \tfrac{\beta}{2}} \end{array}}$$

the sum is zero $= 0$ if $\frac{n\beta}{2} = 0$ provided that $\sin\frac{\beta}{2} \neq 0$

$\therefore \frac{n\beta}{2} = k\pi$ when k is a positive integer.

i.e $\boxed{\beta = \dfrac{2k\pi}{n}}$ provided that $\frac{k}{n}$ is not an integer

$$\text{II} \quad \cos^2\alpha + \cos^2(\alpha+\beta) + \cdots + \cos^2(\alpha+\overline{n-1}\beta) =$$
$$= \tfrac{1}{2}\left[1+\cos 2\alpha\right] + \tfrac{1}{2}\left[1+\cos(2\alpha+2\beta)\right] + \cdots + \tfrac{1}{2}\left[1+\cos(2\alpha+\overline{n-1}\,2\beta)\right]$$
$$= \frac{n}{2} + \frac{1}{2}\left[\cos 2\alpha + \cos(2\alpha+2\beta) + \cdots + \cos(2\alpha+\overline{2n-2}\beta)\right]$$

The sum of the series between square brackets vanishes if
$2\beta = \dfrac{2k\pi}{n}$ provided that $\dfrac{k}{n}$ is not an integer. The
sum of the given series is then equal to $\dfrac{n}{2}$. Similarily

$$\sum_{r=0}^{n-1} \sin^2(\alpha + r\beta) = \frac{n}{2} \qquad \text{when } 2\beta = \frac{2k\pi}{n}$$

$$\text{III)} \quad
\begin{array}{ccccc}
\alpha & \alpha+\beta & \alpha+2\beta & & \alpha+\overline{n-1}\beta \\
\gamma & \gamma+\delta & \gamma+2\delta & & \gamma+\overline{n-1}\delta.
\end{array}$$

consider
for example:

$$\cos\alpha\cos\gamma + \cos(\alpha+\beta)\cos(\gamma+\delta) + \cdots \cos[\alpha+(n-1)\beta]\cos(\gamma+\overline{n-1}\delta) =$$
$$= \tfrac{1}{2}\left[\cos(\alpha+\gamma) + \cos(\alpha-\gamma)\right] + \tfrac{1}{2}\left[\cos(\alpha+\beta+\gamma+\delta) + \cos(\alpha+\beta-\gamma-\delta)\right]$$

$$= \tfrac{1}{2}\left\{\cos(\alpha+\gamma) + \cos(\alpha+\gamma+\beta+\delta) + \cdots\right\} +$$
$$\tfrac{1}{2}\left\{\cos(\alpha-\gamma) + \cos(\alpha-\gamma+\beta-\delta) + \cdots\right\}.$$

The sum vanishes if $\beta + \delta = \dfrac{2k\pi}{n}$
$$\beta - \delta = \frac{2s\pi}{n}$$

112

provided that $\frac{k}{n}$ & $\frac{s}{n}$ are not integers similarly for other products."

11.2. The 12-ordinate scheme for evaluating the coefficients of Fourier transform

The 12 - ordinate scheme

Divide the period 2π into 12 - equal parts each part $=\frac{2\pi}{12}$ $= \frac{\pi}{6}$ first the ordinates $y_0, y_1, y_2, \ldots, y_{11}$ measure these ordinates & construct the following table

$$
\begin{array}{cccccc}
O & \frac{\pi}{6} & \frac{2\pi}{6} & \cdots & \frac{11\pi}{6} \\
y_0 & y_1 & y_2 & & y_{11}
\end{array}
$$

we assume as a form of expansion of the given f^n (graph)

$$y = a_0 + a_1 \cos x + a_2 \cos 2x + \cdots + a_5 \cos 5x + a_6 \cos 6x$$
$$+ b_1 \sin x + b_2 \sin 2x + \cdots + b_5 \sin 5x \circ \quad \underline{\underline{12 \text{ term}}}$$

Substitute the 12-points in this eq^n which contains 12-unknown

$$y_0 = a_0 + a_1 + a_2 + a_3 + \cdots + a_6 \quad\text{——①}$$

$$y_1 = a_1 + a_1 \cos\frac{\pi}{6} + a_2 \cos\frac{2\pi}{6} + \cdots + a_5 \cos\frac{5\pi}{6} + a_6 \cos\frac{6\pi}{6}$$
$$+ b_1 \sin\frac{\pi}{6} + b_2 \sin\frac{2\pi}{6} + \cdots + b_5 \sin\frac{5\pi}{6} \quad\text{——②}$$

$$y_2 = a_0 + a_1 \cos\frac{2\pi}{6} + a_2 \cos\frac{4\pi}{6} + \cdots + a_5 \cos\frac{10\pi}{6} + a_6 \cos\frac{12\pi}{6}$$
$$+ b_1 \sin\frac{2\pi}{6} + b_2 \sin\frac{4\pi}{6} + \cdots + a_5 \sin\frac{10\pi}{6} \quad\text{——③}$$

113

$$y_{11} = a_0 + a_1 \cos 11\frac{\pi}{6} + a_2 \cos \frac{22\pi}{6} + \cdots + a_5 \cos 55\frac{\pi}{6} + a_6 \cos 66\frac{\pi}{6}$$
$$+ b_1 \sin 11\frac{\pi}{6} + b_2 \sin 22\frac{\pi}{6} + \cdots + b_5 \sin 55\frac{\pi}{6} \quad \text{—} \quad (11)$$

we have now to solve these eq$\underline{ns}$ in 12 unknown (12 eq$\underline{n}$)

unknown:- $a_0, a_1, \ldots a_6, b_1, b_2, \ldots b_5$.

① a_0 we add the 12 - eq$\underline{ns}$ we get.

$$\boxed{y_0 + y_1 + y_2 + \cdots + y_{11} = 12 a_0}$$ from Ⅰ

$\beta = \frac{2k\pi}{n}$

$\bullet \ \beta = \frac{\pi}{6} = \frac{2k\pi}{12}$

$\bullet \ n = 12$

all the other sum vanishes (like the previous, all other integrals vanishes) as $\boxed{\beta = \frac{2k\pi}{n}}$

$= k$, $k=1$ integer

in ① a_1's $\frac{\pi}{6} = \frac{2\pi}{12} k = \frac{\pi}{6}$ $k = 1, 2, 3 \ldots$

this show that a_0 is the mean ordinate that we divide by 12

② a_1 - multiply both sides of ① by the co$\underline{eff}^\dagger$ of a_1 which is $\frac{1}{\underline{\underline{\ \ }}}$

$\sim \quad \sim \quad \sim \quad$ ② $\sim \quad \sim \quad \sim \quad a_2 \sim \sim \cos \frac{\pi}{6}$

& so for the other eq$\underline{ns}$ & then add.

from Ⅱ

$$\boxed{y_0 + y_1 \cos \frac{\pi}{6} + y_2 \cos \frac{2\pi}{6} + \cdots + y_{11} \cos 11\frac{\pi}{6} = 6 a_1}$$ $\frac{n}{2}$

③ a_2 multiply both sides of ① by the co$\underline{eff}^\pm$ of a_2 which is unity

$\sim \quad \sim \quad \quad \sim \quad \sim$ ② $\sim \quad \quad \sim \quad \sim a_2 \sim \cos \frac{2\pi}{6}$

& so on & then add.

$$\boxed{y_0 + y_1 \cos \frac{2\pi}{6} + y_2 \cos \frac{4\pi}{6} + \cdots + y_{11} \cos \frac{22\pi}{6} = 6 a_2}$$

$\frac{n}{2}$

114

4 a_5 $\boxed{y_0 + y_1 \cos\dfrac{5\pi}{6} + y_2 \cos 10\dfrac{\pi}{6} + \cdots + y_{11} \cos 55\dfrac{\pi}{6} = 6\,a_5}$ $\underset{12}{6}$

5 a_6 $\boxed{y_0 + y_1 \cos\dfrac{6\pi}{6} + y_2 \cos\dfrac{12\pi}{6} \cdots + y_{11} \cos\dfrac{66\pi}{6} = 12\,a_6}$

$$n = 12$$

$$\beta = \pi \quad = \quad \frac{2\pi k}{n} = 2\pi \frac{6}{12} = \pi$$

$$as \ \cos\frac{6\pi}{6}, \ \cos\frac{12\pi}{6}, \ \cos\frac{180}{6} \cdots \ \cos\frac{66\pi}{6}$$

Note
$$= -1$$
$$\therefore \ a_6 \cos^2\frac{6\pi}{6} + a_6 \cos^2\frac{120}{6} \cdots a_6 \cos^2\frac{66\pi}{6} = 12\,a_6$$

6 b_1 $\boxed{y_1 \sin\dfrac{\pi}{6} + y_2 \sin\dfrac{2\pi}{6} + y \cdots + y_{11} \sin\dfrac{11\pi}{6} = 6\,b_1}$

7 b_5 $\boxed{y \sin 5\dfrac{\pi}{6} + y_2 \sin\dfrac{10\pi}{6} + \cdots + y_{11} \sin\dfrac{55\pi}{6} = 6\,b_5}$

11.3. The circle scheme for evaluating the coefficients of Fourier transform

The circle rule for evaluation of the coeffts

Draw a circle; Divide this circle into 12 equal parts by radial lines. & write y_0 at the Right end of the Horizontal diameter.

take intervales = suffix. for the a's. multiply by cosines & for the b's multiply by sines.

115

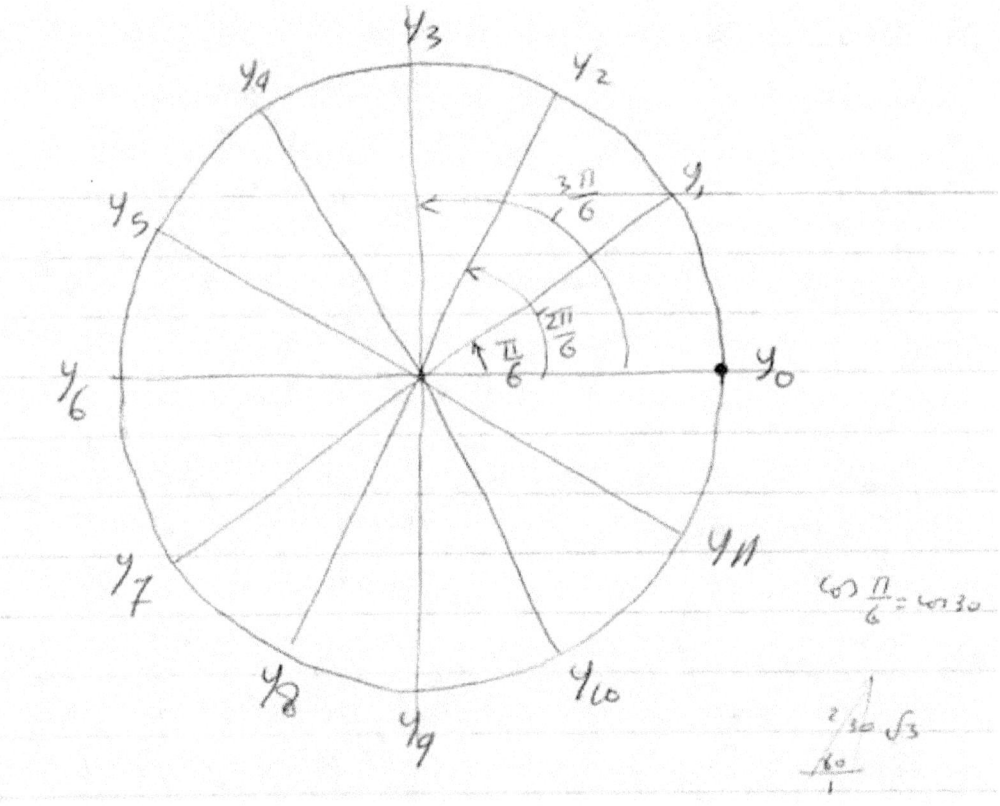

$\cos \frac{\pi}{6} = \cos 30$

$\frac{2}{30} \sqrt{3}$

$\underline{a_1, b_1}$

$$6a_1 = y_0 + y_1 \cos \frac{\pi}{6} + y_2 \cos \frac{2\pi}{6} + \cdots + y_{11} \cos 11 \frac{\pi}{6}$$

$$6a_1 = y_0 - y_6 + \frac{\sqrt{3}}{2}\left[y_1 + y_{11} - y_5 - y_7 \right]$$

$$+ \frac{1}{2}\left[y_2 + y_{10} - y_4 - y_8 \right]$$

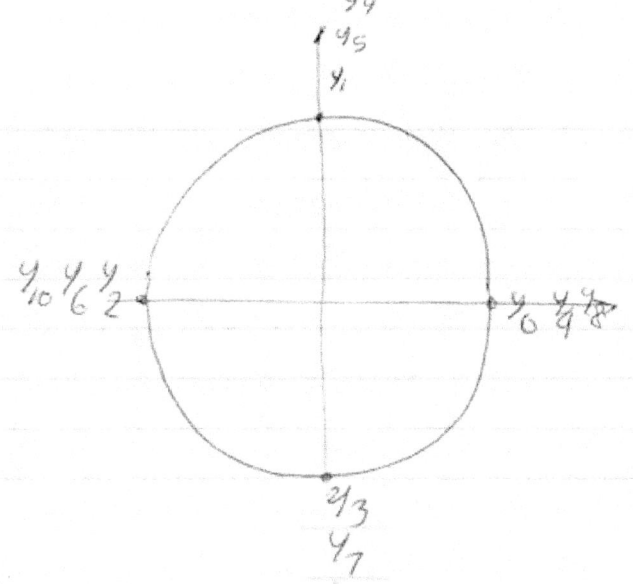

$$6b_1 = \frac{1}{2}\left[y_1 + y_5 - y_{11} - y_7 \right] + \frac{\sqrt{3}}{2}\left[y_2 + y_4 - y_{10} - y_8 \right]$$
$$+ \; y_3 - y_9 \qquad\qquad y_6 = 0$$

a_3, b_3

$$6a_3 = y_0 + y_4 + y_8 - y_{10} - y_6 - y_2$$
$$6b_3 = y_1 + y_5 + y_9 - y_3 - y_7 - y_{11}$$

b_9, a_9

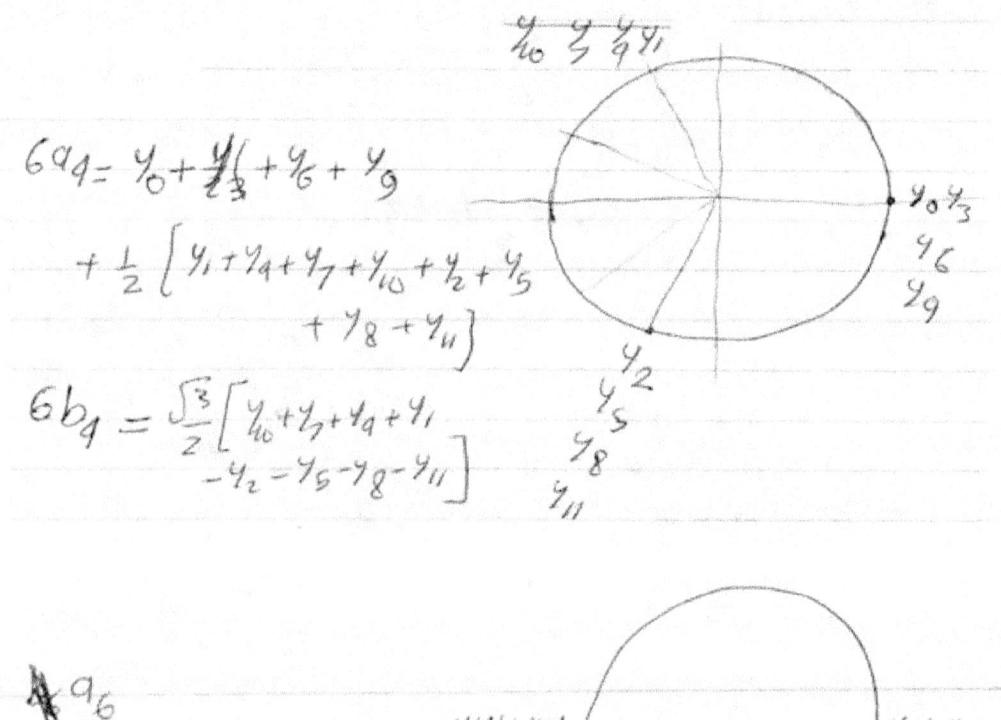

$$6a_9 = y_0 + \frac{y_1}{2}\frac{y_3}{3} + y_6 + y_9$$

$$+ \frac{1}{2}\left[y_1 + y_4 + y_7 + y_{10} + y_2 + y_5 \right.$$
$$\left. + y_8 + y_{11} \right]$$

$$6b_9 = \frac{\sqrt{3}}{2}\left[y_{10} + y_1 + y_4 + y_1 \right.$$
$$\left. - y_2 - y_5 - y_8 - y_{11} \right]$$

a_6

$$12 a_6 = y_0 + y_2 + y_4 + y_6 + y_8 + y_{10}$$
$$- y_1 - y_3 - y_5 - y_7 - y_9 - y_{11}$$

example

complete the following table for odd sines, then find
b_5.

x	$0°$	$30°$	$60°$	$90°$	given only
y	$0°$	4	9	16	

118

Soln

sines: symmetry about the origin $f(-x) = -f(x)$
odd: repeat every π with -ve sign $f(x+\pi) = -f(x)$.

0	30	60	90	120	150	180	210	240	270	300	330	360
0	9	9	15	9	9	0	-9	-9	-15	-9	-9	0

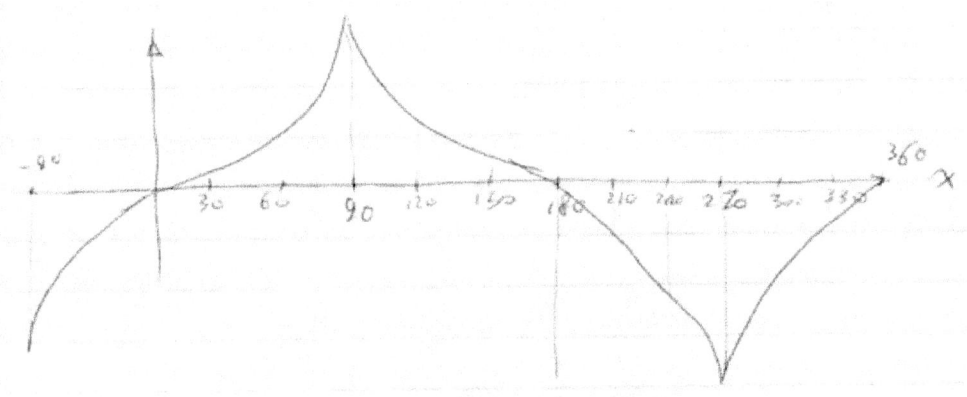

$$6 b_5 = \frac{1}{2}\left[y_5 + y_1 - y_7 - y_{11} \right]$$
$$+ \frac{\sqrt{3}}{2}\left[y_0 + y_8 - y_2 - y_4 \right]$$
$$+ y_3 - y_9$$

Table methods for evaluat^n the coeffs

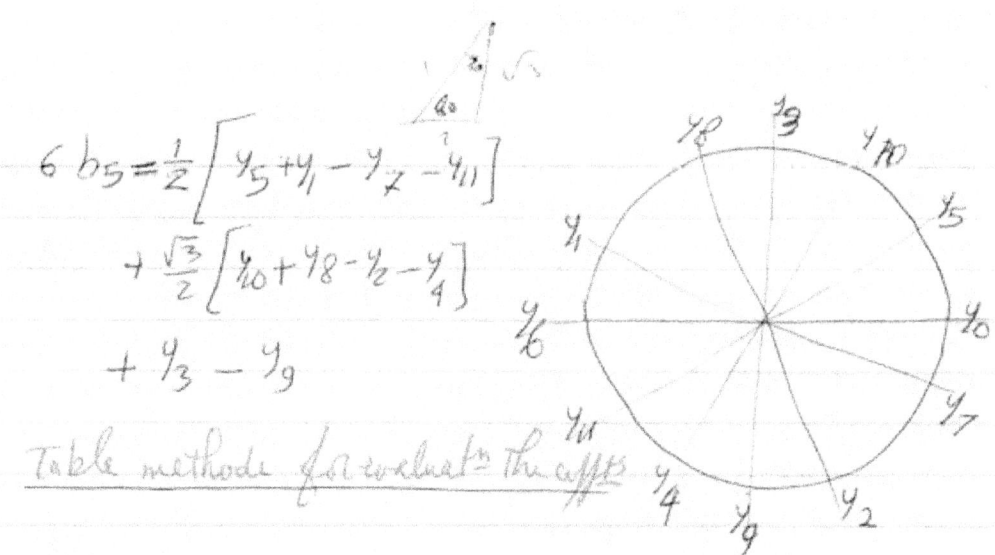

CHAPTER 11

LAPLACE TRANSFORMS

1.1. Integral transforms.

The integral transform f(p) of a given function f(x) in the range (a,b) is defined as follows.

$$\bar{f}(p) = \int_a^b f(x)\, k(p,x)\, dx$$

where k(p,x) is a known function of p and x known as the kernel of the transform. This is on the assumption that the integral exists. If the limits of integration are 0 and ∞ then we have the following special cases :—

(i) If $k(p,x) = e^{-px}$, then

$$\bar{f}(p) = \int_0^\infty f(x)\, e^{-px}\, dx$$

and this is known as the Laplace transform of f(x).

(ii) If $k(p,x) = \sin px$, then

$$\bar{f}(p) = \int_0^\infty f(x)\, \sin px\, dx$$

and this is the infinite Fourier sine transform of f(x).

(iii) If $k(p,x) = \cos px$, then

$$\bar{f}(p) = \int_0^\infty f(x)\, \cos px\, dx$$

and this is the infinite Fourier cosine transform.

(iv) If $k(p,x) = x J_n(px)$ where $J_n(px)$ is the Bessel function of the first kind of order n, then

$$\overline{f}(p) = \int_0^\infty x\, f(x)\, J_n(px)\, dx$$

and this is known as the Hankel transform of f(x).

(v) If $k(p,x) = x^{p-1}$, then

$$\overline{f}(p) = \int_0^\infty f(x)\, x^{p-1} dx$$

and this is known as the Mellia transform.

Such transforms have been widely used to obtain solutions of both ordinary and partial linear differential equations.

The main object of the present work is to discuss in some detail the Laplace transformation and its applications to the solution of differential equations with given initial conditions and in solving boundary — value problems.

1.2. The Laplace transformation.

Let F(t) be a function defined for all positive values of the real variable t. The Laplace transform of F(t) usually denoted
by

$$L\left\{ F(t) \right\} = \int_0^\infty e^{-st}\, F(t)\, dt = f(s)$$

provided that the integral exists. We shall denote the original function by a capital letter and its transform by the corresponding lower case letter. We often call F(t) the object function and f(s) the result function. s may be real or complex but we shall consider first real values of s.

We shall start by finding the Laplace transforms of some of the elementary functions from first principles. Later on simpler methods will be used for the derivation of these transforms and we shall give the limitations on F(t) as well as on the range of s.

I. $L\left\{1\right\}, t > 0$

$$L\left\{1\right\} = \int_0^\infty e^{-st}\, dt = \left[\frac{e^{-st}}{-s}\right]_0^\infty = \frac{1}{s}, \ s > 0$$

$$\therefore \ L\left\{1\right\} = \frac{1}{s}, \ s > 0$$

II. $L\left\{e^{at}\right\}, t > 0$

$$L\left\{e^{at}\right\} = \int_0^\infty e^{-st} \cdot e^{at}\, dt = \int_0^\infty e^{-(s-a)t}\, dt$$

$$= \left[\frac{e^{-(s-a)t}}{-(s-a)}\right]_0^\infty = \frac{1}{s-a}, \ s > a$$

$$\therefore \ L\left\{e^{at}\right\} = \frac{1}{s-a}, \ s > a$$

III. $L\left\{t^k\right\}$ where $k > -1$

$$L\left\{t^k\right\} = \int_0^\infty e^{-st}\, t^k\, dt$$

Put $st = x$ where $s > 0$ $\therefore t = \frac{x}{s}$, $dt = \frac{dx}{s}$

122

$$\therefore L\left\{ t^k \right\} = \int_0^\infty e^{-x} \frac{x^k}{s^k} \frac{dx}{s} = \frac{1}{s^{k+1}} \int_0^\infty e^{-x} x^k \, dx$$

$$= \frac{\Gamma(k+1)}{s^{k+1}}$$

Note that the condition $k > -1$ is necessary for the convergence of the integral.

$$\therefore L\left\{ t^k \right\} = \frac{\Gamma(k+1)}{s^{k+1}}, \quad k > -1, \; s > 0$$

Corollary. 1.   If k is a positive integer equal to n, then

$$L\left\{ t^n \right\} = \frac{\Gamma(n+1)}{s^{n+1}} = \frac{n!}{s^{n+1}}, \quad s > 0$$

Thus $L\left\{ t \right\} = \frac{1}{s^2}$, $L\left\{ t^2 \right\} = \frac{2!}{s^3}$, $L\left\{ t^3 \right\} = \frac{3!}{s^4}$, ...

Corollary. 2. If $k = -\frac{1}{2}$, then

$$L\left\{ t^{-\frac{1}{2}} \right\} = \frac{\Gamma\left(\frac{1}{2}\right)}{s^{\frac{1}{2}}} = \frac{\sqrt{\pi}}{\sqrt{s}} = \sqrt{\frac{\pi}{s}}, \quad s > 0$$

IV. $L\left\{ \sin kt \right\} = \int_0^\infty e^{-st} \sin kt \, dt$

$$= \left[-\frac{(s \sin kt + k \cos kt) e^{-st}}{s^2 + k^2} \right]_0^\infty = \frac{k}{s^2 + k^2}, \quad s > 0$$

$$L\left\{\cos kt\right\} = \int_0^\infty e^{-st} \cos kt \, dt$$

$$= \left[\frac{(k \sin kt - s \cos kt) \, e^{-st}}{s^2 + k^2}\right]_0^\infty = \frac{s}{s^2 + k^2}, \, s > 0$$

1.3. The Laplace transformation of the sum of two functions

Let A and B be two arbitrary constants, then

$$L\left\{A \, F(t) + B \, G(t)\right\} = \int_0^\infty e^{-st}\left\{A \, F(t) + B \, G(t)\right\} dt$$

$$= A \int_0^\infty e^{-st} F(t) \, dt + B \int_0^\infty e^{-st} G(t) \, dt$$

$$= A \, L\left\{F(t)\right\} + B \, L\left\{G(t)\right\}$$

This means that the Laplace transform of a linear combination of two functions is the same linear combination of the transforms of these functions.

$$\text{e. g.} \quad L\left\{\cosh kt\right\} = L\left\{\frac{e^{kt} + e^{-kt}}{2}\right\}$$

$$= \tfrac{1}{2}\left[\frac{1}{s-k} + \frac{1}{s+k}\right] = \frac{s}{s^2-k^2}, \quad s > |k|$$

$$\text{Similarly} \quad L\left\{\sinh kt\right\} = \frac{k}{s^2 - k^2}, \quad s > |k|$$

1.4. Sectionally or piecewise continuous functions

A function is said to be sectionally continuous or piecewise continuous in an interval $a \leqslant t \leqslant b$ if this interval can be subdivided into a finite number of intervals in each of which the function is continuous and has finite right and left hand limits. It follows

124

from the definition that sectionally continuous functions include two main classes of functions:
(i) continuous functions,
(ii) discontinuous functions in which the discontinuity consists of finite jumps,

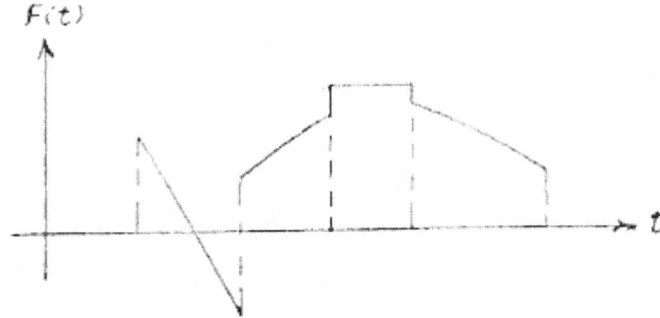

It is worthy of notice that sectionally continuous functions are integrable functions, the integral being the sum of the integrals of the continuous functions over the subintervals.

Example 1. Find the Laplace transform of the function F(t) defined by

$$F(t) = F_0 \qquad 0 < t < t_0$$
$$F(t) = 0 \qquad t > t_0$$

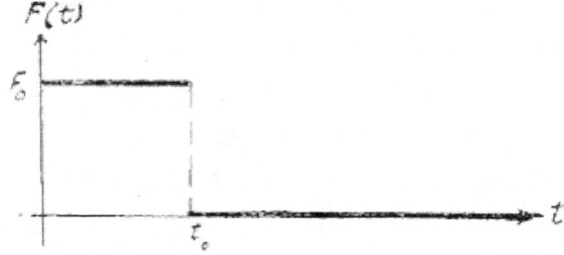

Solution:

$$L\left\{F(t)\right\} = \int_0^\infty e^{-st}\ F(t)\ dt = \int_0^{t_0} e^{-st}\ F_0\ dt$$

$$= F_0\left[-\frac{e^{-st}}{-s}\right]_0^{t_0} = \frac{F_0}{s}\left(1 - e^{-t_0 s}\right)$$

125

Example 2. Find the Laplace transform of H(t) where H(t) = t at o<t<4 and H(t) = 5 at t > 4

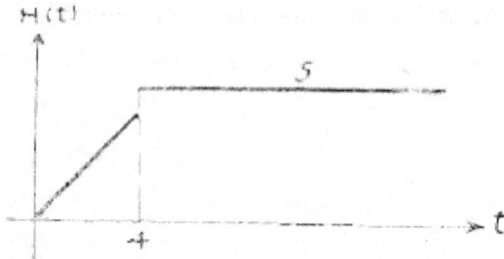

Solution:

$$L\left\{H(t)\right\} = \int_0^\infty e^{-st}\; H(t)\; dt$$

$$= \int_0^4 e^{-st}\; t\; dt + \int_4^\infty e^{-st} \times 5\; dt$$

$$= \left[-\frac{t}{s}\; e^{-st} - \frac{1}{s^2}\; e^{-st} \right]_0^4 + \left[-\frac{5}{s}\; e^{-st} \right]_4^\infty$$

$$= \frac{1}{s^2} + \frac{e^{-4s}}{s} - \frac{e^{-4s}}{s^2}$$

1.5. Functions of exponential order
If two positive constants M and a exist such that for all

$$t > T, \; \left| e^{-at}\; F(t) \right| < M \quad \text{or} \quad \left| F(t) \right| < Me^{at}\;,$$

then we say that F(t) is a function of exponential order a as $t \longrightarrow \infty$ and write F(t) is $O\left(e^{at}\right)$. This actually means that functions of exponential order cannot grow in absolute value more rapidly than Me^{at} as t increases.

Example 1. Let F(t) = C where C is a constant. Choose a positive constant M such that M > | C |

126

$$\therefore \quad |C| < M\,e^{at} \text{ as } t \longrightarrow \infty \text{ for all } a \geq 0$$

$$\therefore \quad C \text{ is } O\left(e^{at}\right) \text{ where } a \geq 0$$

Similarly for all bounded functions e.g. sin kt or cos kt.
Thus sin kt and cos kt are

$$O\left(e^{at}\right) \text{ where } a \geq 0.$$

Example 2. Let $F(t) = t^n$ to where n is a positive integer.

$$\frac{t^n}{e^{at}} \longrightarrow o \text{ as } t \longrightarrow \infty \text{ for all } a > 0$$

$$\therefore \quad \left| \frac{t^n}{e^{at}} - o \right| < M \text{ as } t \longrightarrow \infty.$$

$$\text{i.e.} \quad \left| t^n \right| < M\,e^{at} \text{ as } t \longrightarrow \infty \text{ for all } a > 0.$$

$$\therefore \quad t^n \text{ is } O\left(e^{at}\right) \text{ where } a > 0.$$

Example 3. Let

$$F(t) = e^{t^3}$$

$$\therefore \quad \left| e^{-at} \cdot e^{t^3} \right| = e^{t^3 - at}$$

can be made larger than any given constant by increasing t. Hence e^{t^3} is not of exponential order.

1.6. Sufficient conditions for the existence of the Laplace transform

We shall now show that the Laplace transform of F(t) exists when:

(i) F(t) is sectionally continuous in every finite interval in the range $t \geqslant 0$, (ii) F(t) is $O(e^{at})$ as $t \to \infty$,

i.e $|F(t)| < Me^{at}$ for all $t > T$ (say).

For $$\int_0^\infty e^{-st} F(t)\, dt = \int_0^T e^{-st} F(t)\, dt + \int_T^\infty e^{-st} F(t)\, dt$$

The first integral on the R.H.S. exists since F(t) is sectionally continuous in every finite interval $0 \leqslant t \leqslant T$. The second integral on the right also exists since F(t) is $O(e^{at})$ for $t > T$. This follows from the fact that

$$\left| \int_T^\infty e^{-st} F(t)\, dt \right| \leqslant \int_T^\infty \left| e^{-st} F(t) \right| dt$$

$$< \int_T^\infty e^{-st} Me^{at}\, dt$$

$$= M \int_T^\infty e^{-(s-a)t}\, dt = M \left[\frac{e^{-(s-a)t}}{-(s-a)} \right]_T^\infty$$

$$= \frac{Me^{-(s-a)T}}{s-a}, \quad s > a$$

$$< \frac{M}{s-a}$$

Hence the Laplace integral $\int_0^\infty e^{-st} F(t)\, dt$ exists if $s > a$.

128

It should be noticed that the above conditions though sufficient are not necessary. They are however simple to apply in most practical problems. If the above conditions are not satisfied the Laplace transform may or may not exist. Thus for example

$$L\left\{t^{-\frac{1}{2}}\right\} = \sqrt{\frac{\pi}{s}}, \quad s > 0$$

as we have already shown in 1.2, but $F(t) = t^{-\frac{1}{2}}$ does not satisfy the above conditions.

1.7. Null functions

A null function N(t) is a function of t such that for all t > 0

$$\int_0^t N(\tau)\, d\tau = 0$$

e.g. the function

F(t) = 2, t = 1
F(t) = 1, t = 2
F(t) = -1, t ---, 3
F(t) = 0 otherwise
is a null function. In general any function which is zero at all but a countable set of points is a null function. It is evident that the Laplace transform of a null function is zero.

1.8. Inverse Laplace transforms

If L { F(t)} = f(s) , then the inverse Laplace transform is often written

$$L^{-1}\left\{f(s)\right\} = F(t)$$

i.e the inverse Laplace transform of f(s) is F(t). Now the Laplace transform of a null function N (t) is zero. Hence if

L {F(t)} = f(s) then L {F(t) + N(t)} = f(s) .

Thus we can have two different functions with the same Laplace transform. Hence if we allow null functions the inverse transform is not unique, but if we do not allow null functions, then it has been shown by Lerch that if we restrict ourselves to functions of exponential order for t > T and which are sectionally continuous in every finite interval $0 \leqslant t \leqslant T$

Then

$$L^{-1} \left\{ f(s) \right\} = F(t)$$

is unique, We shall now write the previous transforms which we have already obtained in the inverse form.

$$L^{-1} \left\{ \frac{1}{s} \right\} = 1 , \quad L^{-1} \left\{ \frac{1}{s^n} \right\} = \frac{t^{n-1}}{(n-1)!} \quad (n \text{ is a } + \text{ve integer})$$

$$L^{-1} \left\{ \frac{1}{s-a} \right\} = e^{at} , \quad L^{-1} \left\{ \frac{1}{s+a} \right\} = e^{-at}$$

$$L^{-1} \left\{ \frac{1}{s^2+k^2} \right\} = \frac{1}{k} \sin kt , \quad L^{-1} \left\{ \frac{s}{s^2+k^2} \right\} = \cos kt$$

$$L^{-1} \left\{ \frac{1}{s^2-k^2} \right\} = \frac{1}{k} \sinh kt , \quad L^{-1} \left\{ \frac{s}{s^2-k^2} \right\} = \cosh kt$$

1.9. The inverse Laplace transformation of the sum of two functions

We have shown that

$$L \left\{ A F(t) + B G(t) \right\} = A f(s) + B g(s)$$

$$\therefore L^{-1} \left\{ A f(s) + B g(s) \right\} = A F(t) + B G(t)$$

$$= A L^{-1} \left\{ f(s) \right\} + B L^{-1} \left\{ g(s) \right\}$$

Example.

$$L^{-1} \left\{ \frac{3}{s+1} + \frac{2s}{s^2+25} - \frac{4}{s^2+9} \right\}$$

$$= L^{-1} \left\{ \frac{3}{s+1} \right\} + L^{-1} \left\{ \frac{2s}{s^2+25} \right\} - L^{-1} \left\{ \frac{4}{s^2+9} \right\}$$

$$= 3e^{-t} + 2\cos 5t - \frac{4}{3} \sin 3t$$

1.10. Transforms of derivatives

Let F (t) be a function of order e^{at} as $t \longrightarrow \infty$ and suppose that F(t) is continuous with a sectionally continuous derivative F'(t) in every finite interval $0 \leqslant t \leqslant T$. Then provided s > a, the Laplace transform of F'(t) exists and is given by

$$L \left\{ F'(t) \right\} = s L \left\{ F(t) \right\} - F(o)$$
$$= s\, f(s) - F(o)$$

where F(o) owing to the continuity of F(t) is the same as F(o+), i.e., the same as the limiting value of F(t) when 1 approaches 0 through positive values.

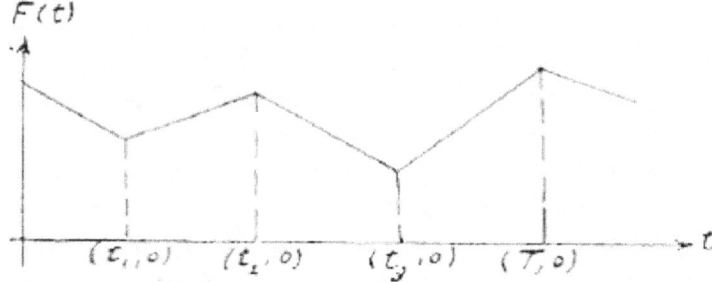

The figure shows a continuous function with a sectionally continuous derivative in the interval $0 \leqslant t \leqslant T$ and we have

$$L \left\{ F'(t) \right\} = \lim_{T \longrightarrow \infty} \int_0^T e^{-st}\, F'(t)\, dt$$

Now $\displaystyle\int_0^T e^{-st}\, F'(t)\, dt$

$$= \int_0^{t_1} e^{-st}\, F'(t)\, dt + \int_{t_1}^{t_2} e^{-st}\, F'(t)\, dt + .. + \int_{t_n}^T e^{-st}\, F'(t)\, dt$$

$$= \int_o^{t_1} e^{-st} \, d\,F(t) + \int_{t_1}^{t_2} e^{-st} \, d\,F(t) + \cdots + \int_{t_n}^{T} e^{-st} \, d\,F(t)$$

Integrating by parts we get

$$\int_0^T e^{-st} \, F'(t) \, dt = \left[e^{-st} \, F(t) \right]_0^{t_1} + \left[e^{-st} \, F(t) \right]_{t_1}^{t_2} + \cdots$$

$$+ \left[e^{-st} \, F(t) \right]_{t_n}^{T} + s \int_0^T e^{-st} \, F(t) \, dt$$

Now, owing to the continuity of $F(t)$, $F(t_1 - 0) = F(t_1 + 0)$

i.e. the limiting value of $F(t)$ as $t \longrightarrow t1$ is the same whether we approach $t1$ from the right or left and similarly for $t\,2, t\,3$, ,..., $t\,n$.

$$\therefore \int_0^T e^{-st} \, F'(t) \, dt = -\,F(o) + e^{-sT}\,F(T) + s \int_0^T e^{-st} \, F(t) \, dt$$

Now, since $F(t)$ is $O(e^{at})$, then $|F(t)| < M\,e^{at}$ for large t and consequently

$$|e^{-sT}\,F(T)| < M\,e^{-(s-a)T} \longrightarrow 0 \text{ as } T \longrightarrow \infty \quad (s > a)$$

Also $\int_0^T e^{-st} \, F(t) \, dt \longrightarrow L \left\{ F(t) \right\}$ as $T \longrightarrow \infty$

$$\therefore \int_0^\infty e^{-st} \, F'(t) \, dt = -\,F(0) + s\,L \left\{ F(t) \right\}$$

132

i.e. $L\left\{F'(t)\right\} = s\,f(s) - F(o)$

The transform of the second derivative F"(t) can be obtained in a similar manner by applying the above theorem to F'(t). Assuming F(t) to be continuous and of exponential order and F"(t) to be sectionally continuous, then

$$L\left\{F''(t)\right\} = s\,L\left\{F'(t)\right\} - F'(o)$$

$$= s\left\{s\,f(s) - F(o)\right\} - F'(o)$$

$$= s^2\,f(s) - s\,F(o) - F'(o)$$

In general suppose that F(t) together with its derivatives

$F'(t)$, $F''(t)$, ..., $F^{(n-1)}(t)$ be of order e^{at} as $t \longrightarrow \infty$

and that F(t) has a continuous derivative $F^{(n-1)}(t)$ and a sectionally continuous derivative $F^{(n)}t$ in every finite interval $0 \leqslant t \leqslant T$, then

$$L\left\{F^{(n)}(t)\right\} = s^n\,f(s) - s^{n-1}\,F(0) - s^{n-2}\,F'(0) - \ldots - F^{(n-1)}(0)$$

Example 1. Find $L\{\sinh kt\}$

Solution:

Put $F(t) = \sinh kt$, then $F'(t) = k\cosh kt$, $F''(t) = k^2 \sinh kt$

$F(o) = o$, $F'(o) = k$

$L\left\{F''(t)\right\} = s^2\,f(s) - s\,F(o) - F'(o)$

$\therefore L\left\{k^2 \sinh kt\right\} = s^2\,f(s) - k$

i.e $k^2\,L\left\{\sinh kt\right\} = s^2\,L\left\{\sinh kt\right\} - k$

$\therefore L\left\{\sinh kt\right\} = \dfrac{k}{s^2 - k^2}$

Example 2. Find $L\left\{t^n\right\}$ where n is a positive integer.

Solution:

Put $F(t) = t^n$, then $F'(t) = n\,t^{n-1}$, $F''(t) = n(n-1)t^{n-2}$ ...

$$F^{(n)}(t) = n!$$

$$\therefore \quad F(o) = F'(o) = F''(o) = .. = F^{(n-1)}(o) = o$$

$$L\left\{F^{(n)}(t)\right\} = s^n\, f(s) - s^{n-1}\, F(o) - s^{n-2}\, F'(o) ... - F^{(n-1)}(o)$$

$$\therefore L\left\{n!\right\} = s^n\, L\left\{t^n\right\}$$

$$\therefore \frac{n!}{s} = s^n\, L\left\{t^n\right\} \quad \text{i.e} \quad L\left\{t^n\right\} = \frac{n!}{s^{n+1}}$$

Example 3. For the function

$$F(t) = t + 1 \quad 0 \leqslant t \leqslant 2$$
$$= 3 \qquad\quad t > 2$$

draw the graphs of F(t) and F'(t) and find L { F'(t) }

Solution:

$$L\left\{F'(t)\right\} = \frac{1}{s}\left(1 - e^{-2s}\right) , s > 0$$

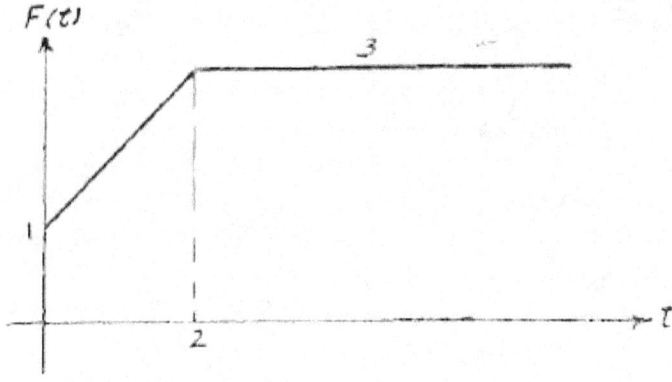

134

1.11. Transforms of integrals.

If F(t) be sectionally continuous and of exponential order then

$$L\left\{\int_0^t F(\tau)\ d\tau\right\} = \frac{1}{s}\ f(s)\quad \text{where}\quad f(s) = L\left\{F(t)\right\}$$

0

Put

$$G(t) = \int_0^t F(\tau)\ d\tau$$

then G(t) is continuous and of exponential order and G(o) = o. Since G'(t) = F(t) then transforming both sides we get

$$s\,L\left\{G(t)\right\} - G(o) = f(s)$$

$$\therefore L\left\{G(t)\right\} = \frac{1}{s}\ f(s)$$

$$\text{i.e } L\left\{\int_0^t F(\tau)\ d\tau\right\} = \frac{1}{s}\ f(s)$$

Hence we have the following result:

The division of the transform of a function by s corresponds to the integration of that function between o and t. The rule can be extended to a repeated division by s.

Thus the division of the transform of F(t) by s^n corresponds to a repeated integration of F(t) n times from o to t.

135

Example.   We know that

$$L\left\{\sin kt\right\} = \frac{k}{s^2 + k^2}, \ s > 0$$

$$\therefore L\left\{\int_0^t \sin k\tau \ d\tau\right\} = \frac{k}{s\ (s^2+k^2)}$$

$$\therefore \ L\left\{\left[-\frac{\cos k\tau}{k}\right]_0^t\right\} = \frac{k}{s\ (s^2 + k^2)}$$

i.e $L\left\{\frac{1}{k_2}\ (1-\cos kt)\right\} = \dfrac{1}{s\ (s^2 + k^2)}$

By integrating again we get:

$$L\left\{\frac{1}{k^3}\ (kt - \sin kt)\right\} = \frac{1}{s^2\ (s^2+k^2)}$$

Hence we have the following two results:

$$L^{-1}\left\{\frac{1}{s\ (s^2 + k^2)}\right\} = \frac{1}{k^2}\ (1 - \cos kt)$$

$$L^{-1}\left\{\frac{1}{s^2\ (s^2 + k^2)}\right\} = \frac{1}{k^3}\ (kt - \sin kt)$$

1.12. The first shift theorem of multiplying the object function by e^{at}

Let F(t) be a sectionally continuous function of order e^{at}. Then

$$\text{Then } L\left\{F(t)\right\} = \int_0^\infty e^{-st}\ F(t)\ dt = f(s), \ s > a$$

$$\therefore L\left\{e^{at}\ F(t)\right\} = \int_0^\infty e^{-st}\ .\ e^{at}\ F(t)\ dt$$

136

$$= \int_0^\infty e^{-(s-a)t} F(t)\, dt$$

$$= f(s-a), \quad s-a > \alpha \quad \text{i.e } s > a + \alpha$$

Hence we have the following theorem on substitution:

The change of the variable s in the transform f(s) of F(t) into s - a corresponds to the multiplication of F(t) by e^{at}

Since f(s-a) is a shift of f(s) to the right a distance a, the above theorem is known as the first shift theorem.

e.g. $\quad L \left\{ e^{at} \sin kt \right\} = \dfrac{k}{(s-a)^2 + k^2}, \quad s > a,$

$\quad L \left\{ e^{-at} \cos kt \right\} = \dfrac{s + a}{(s+a)^2 + k^2}, \quad s > -a,$

$\quad L^{-1} \left\{ \dfrac{1}{(s-a)^n} \right\} = e^{at} \dfrac{t^{n-1}}{(n-1)!}$

$\quad L \left\{ e^{3t} \cosh 4t \right\} = \dfrac{s - 3}{(s-3)^2 - 16} = \dfrac{s - 3}{s^2 - 6s - 7}$

Example. Prove that

$$L \left\{ e^{at} t^{-\frac{1}{2}} (1 + 2at) \right\} = \sqrt{\pi} \, \dfrac{s}{(s-a)^{3/2}}$$

Solution:

$$L \left\{ e^{at} t^{-1/2} (1 + 2at) \right\} = L \left\{ e^{at} \left(t^{-1/2} + 2at^{1/2} \right) \right\}$$

$$= L \left\{ e^{at} t^{-1/2} \right\} + 2aL \left\{ e^{at} t^{1/2} \right\}$$

Since $L\left\{t^k\right\} = \dfrac{\Gamma(k+1)}{s^{k+1}}$, $k > -1$

then $L\left\{t^{-1/2}\right\} = \sqrt{\dfrac{\pi}{s}}$,

$$L\left\{t^{1/2}\right\} = \frac{\Gamma(3/2)}{s^{3/2}} = \frac{\frac{1}{2}\,\Gamma(1/2)}{s^{3/2}} = \frac{\sqrt{\pi}}{2\,s^{3/2}}$$

$$\therefore\ L\left\{e^{at}\,t^{-\frac{1}{2}}\right\} = \frac{\sqrt{\pi}}{(s-a)^{1/2}}$$

$$L\left\{e^{at}\,t^{\frac{1}{2}}\right\} = \frac{\sqrt{\pi}}{2\,(s-a)^{3/2}}$$

$$\therefore\ L\left\{e^{at}\,t^{-\frac{1}{2}}(1+2at)\right\} = \frac{\sqrt{\pi}}{(s-a)^{\frac{1}{2}}} + \frac{2a\,\sqrt{\pi}}{2\,(s-a)^{3/2}}$$

$$= \frac{\sqrt{\pi}\ s}{(s-a)^{3/2}}$$

1.13. Two useful Transforms

(i) We know that

$$L\left\{\cos kt\right\} = \int_0^\infty e^{-st}\cos kt\ dt = \frac{s}{s^2+k^2},\ s>0$$

Differentiate w.r.t. k

$$\frac{\partial}{\partial k}\int_0^\infty e^{-st}\cos kt\ dt = \int_0^\infty \frac{\partial}{\partial k}(e^{-st}\cos kt)\ dt$$

$$= -\int_0^\infty e^{-st}\ t\ \sin kt\ dt = \frac{-2ks}{(s^2+k^2)^2}$$

$$\therefore L\left\{t\ \sin\ kt\right\} = \frac{2ks}{(s^2+k^2)^2},\ s > 0$$

138

or $L^{-1}\left\{\dfrac{s}{(s^2+k^2)^2}\right\} = \dfrac{1}{2k}\, t\,\sin kt$

(ii) Dividing the transform by s which corresponds to the integration' of the object function between o and t we get

$$L^{-1}\left\{\dfrac{1}{(s^2+k^2)^2}\right\} = \dfrac{1}{2k}\int_0^t \tau \sin k\tau\, d\tau$$

$$= -\dfrac{1}{2k^2}\int_0^t \tau\, d\cos k\tau$$

$$= -\dfrac{1}{2k^2}\left[\tau \cos k\tau - \dfrac{\sin k\tau}{k}\right]_0^t$$

$$= \dfrac{1}{2k^3}\left(\sin kt - kt \cos kt\right)$$

1.14. The multiplication of the variable by a positive constant

Suppose that $L\{F(t)\} = f(s)$, for $s > a$, and let a be a positive constant, then

$$L\left\{F(at)\right\} = \int_0^\infty e^{-st}\, F(at)\, dt.\ \text{Put}\ at = \tau,\ \text{then}$$

$$L\left\{F(at)\right\} = \dfrac{1}{a}\int_0^\infty e^{-\frac{s\tau}{a}}\, F(\tau)\, d\tau = \dfrac{1}{a}\, f\left(\dfrac{s}{a}\right),\ \dfrac{s}{a} > a$$

$$\therefore\ L\left\{F(at)\right\} = \dfrac{1}{a}\, f\left(\dfrac{s}{a}\right)\ \text{where}\ s > a_\alpha,\ a > 0.$$

This formula can also be written in the form

$$L^{-1}\left\{f(cs)\right\} = \dfrac{1}{c}\, F\left(\dfrac{t}{c}\right),\ c > 0.$$

e.g. $L\left\{\sin t\right\} = \dfrac{1}{s^2+1}$

139

$$\therefore \quad L\left\{ \sin kt \right\} = \frac{1}{k} \cdot \frac{1}{\dfrac{s^2}{k^2}+1} = \frac{k}{s^2+k^2}, \ s>0.$$

1.15. Determination of the inverse transforms by the aid of partial fractions.

A proper fraction $\dfrac{p(s)}{q(s)}$ in which the degree of the numerator is less than that of the denominator of the form

$$\frac{p(s)}{(s-a)\,(s-b)^r\,(s^2+cs+d)\,(s^2+ls+m)^n}$$

can be resolved in a series of partial fractious in the form

$$\frac{p(s)}{q(s)} = \frac{A}{s-a} + \left[\frac{B_1}{(s-b)^r} + \frac{B_2}{(s-b)^{r-1}} + \cdots + \frac{Br}{s-b} \right]$$

$$+ \frac{Cs+D}{s^2+cs+d} + \left[\frac{E_1s+F_1}{(s^2+ls+m)^n} + \frac{E_2s+F_2}{(s^2+ls+m)^{n-1}} \right.$$

$$\left. + \cdots + \frac{E_n s + F_n}{s^2+ls+m} \right]$$

On clearing fractions on both sides we obtain an identity from which the unknown constants can be obtained either by substituting zeros of the denominator in the identity or equating equal powers on both sides.

The following are examples on finding the inverse transform by resolving the transform into its partial fractions.

Example 1.

Find $L^{-1}\left\{ \dfrac{s^2+3}{(s+1)\,(s-2)\,(s-4)} \right\}$

Solution:

$$\frac{s^2+3}{(s+1)\,(s-2)\,(s-4)} = \frac{4}{15\,(s+1)} - \frac{7}{6\,(s-2)} + \frac{19}{10\,(s-4)}$$

$$\therefore \quad L^{-1}\left\{\frac{s^2+3}{(s+1)(s-2)(s-4)}\right\} = \frac{4}{15}\,e^{-t} - \frac{7}{6}\,e^{2t} + \frac{19}{10}\,e^{4t}$$

Example 2.

Find $L^{-1}\left\{\dfrac{s+2}{s^3\,(s-1)^2}\right\}$

Solution:

$$\frac{s+2}{s^3(s-1)^2} = \frac{2}{s^3} + \frac{5}{s^2} + \frac{8}{s} + \frac{3}{(s-1)^2} - \frac{8}{s-1}$$

$$\therefore \quad L^{-1}\left\{\frac{s+2}{s^3(s-1)^2}\right\} = t^2 + 5t + 8 + e^t\,(3t-8)$$

Example 3.

Find $L^{-1}\left\{\dfrac{s^2+1}{(s-1)(s-2)^2}\right\}$

Solution:

Let $\dfrac{s^2+1}{(s-1)(s-2)^2} = \dfrac{A}{s-1} + \dfrac{B}{(s-2)^2} + \dfrac{C}{s-2}$ ... (1)

Multiply both sides by s-1 and let $s \rightarrow 1$

$$\lim_{s \to 1} \frac{s^2+1}{(s-2)^2} = A \quad \therefore \quad A = 2$$

Multiply both sides of (1) by $(s-2)^2$ and let s $\rightarrow$ 2

$$\lim_{s \to 2} \frac{s^2+1}{s--1} = B \quad \therefore \quad B = 5$$

To find C multiply both sides of (1) by s and let s $\rightarrow$ ∞

$$\lim_{s \to \infty} \frac{s^3+s}{(s-1)(s-2)^2} = \lim_{s \to \infty} \frac{As}{s-1} + \lim_{s \to \infty} \frac{Bs}{(s-2)^2}$$

$$+ \lim_{s \to \infty} \frac{Cs}{s-2}$$

141

$$\therefore \quad 1 = A + C = 2 + C \quad \therefore \quad C = -1$$

$$\therefore \quad \frac{s^2 + 1}{(s-1)(s-2)^2} = \frac{2}{s-1} + \frac{5}{(s-2)^2} - \frac{1}{s-2}$$

$$\therefore \quad L^{-1}\left\{\frac{s^2+1}{(s-1)(s-2)^2}\right\} = 2e^t + 5t\, e^{2t} - e^{2t}$$

Example 4.

$$\text{Find } L^{-1}\left\{\frac{3s^2 + 9s + 16}{(s+1)(s^2+4s+13)}\right\}$$

Solution:

$$\frac{3s^2 + 9s + 16}{(s+1)(s^2+4s+13)} = \frac{1}{s+1} + \frac{2s+3}{s^2+4s+13}$$

$$= \frac{1}{s+1} + \frac{2(s+2) - 1}{(s+2)^2 + 9}$$

$$= \frac{1}{s+1} + \frac{2(s+2)}{(s+2)^2 + 9} - \frac{1}{(s+2)^2+9}$$

$$\therefore \quad L^{-1}\left\{\frac{3s^2 + 9s + 16}{(s+1)(s^2+4s+13)}\right\} = e^{-t} + 2e^{-2t}\cos 3t$$

$$- \tfrac{1}{3}\, e^{-2t} \sin 3t$$

Example 5.

$$\text{Find } L^{-1}\left\{\frac{2s+5}{(s^2 + 6s + 25)^2}\right\}$$

Solution:

$$\frac{2s+5}{(s^2 + 6s + 25)^2} = \frac{2(s+3) - 1}{[(s+3)^2 + 16]^2}$$

$$= \frac{2(s+3)}{[(s+3)^2 + 16]^2} - \frac{1}{[(s+3)^2 + 16]^2}$$

142

Now $L^{-1}\left\{\dfrac{s}{(s^2 + k^2)^2}\right\} = \dfrac{1}{2k}\, t\, \sin\, kt$

and $L^{-1}\left\{\dfrac{1}{(s^2 + k^2)^2}\right\} = \dfrac{1}{2k^3}\, (\sin\, kt - kt\, \cos\, kt)$

$\therefore\; L^{-1}\left\{\dfrac{2s + 5}{(s^2 + 6s + 25)^2}\right\}$

$$= \dfrac{2}{8}\, e^{-3t}\, t\, \sin\, 4t - \dfrac{e^{-3t}}{128}\, (\sin\, 4t - 4t\, \cos\, 4t)$$

$$= \dfrac{e^{-3t}}{128}\, (32\, t\, \sin\, 4t - \sin\, 4t + 4t\, \cos\, 4t)$$

Example 6.

Find $L^{-1}\left\{\dfrac{1}{(s^2 + a^2)(s^2 + b^2)}\right\}$, $\quad a^2 \neq b^2$

Solution:

$$\dfrac{1}{(s^2 + a^2)(s^2 + b^2)} = \dfrac{1}{b^2 - a^2}\left(\dfrac{1}{s^2 + a^2} - \dfrac{1}{s^2 + b^2}\right)$$

$\therefore\; L^{-1}\left\{\dfrac{1}{(s^2 + a^2)(s^2 + b^2)}\right\} = \dfrac{1}{b^2 - a^2}\left(\dfrac{1}{a}\, \sin\, at - \dfrac{1}{b}\, \sin\, bt\right)$

Example 7.

$L^{-1}\left\{\dfrac{s}{(s^2 + a^2)(s^2 + b^2)}\right\} = L^{-1}\left\{\dfrac{1}{b^2 - a^2}\left(\dfrac{s}{s^2 + a^2} - \dfrac{s}{s^2 + b^2}\right)\right\}$

$$= \dfrac{1}{b^2 - a^2}\, (\cos\, at - \cos\, bt)$$

Example 8.

$L^{-1}\left\{\dfrac{s^2}{(s^2 + a^2)(s^2 + b^2)}\right\} = L^{-1}\left\{\dfrac{1}{b^2 - a^2}\left(\dfrac{s^2}{s^2 + a^2} - \dfrac{s^2}{s^2 + b^2}\right)\right\}$

$$= L^{-1}\left\{\dfrac{1}{b^2 - a^2}\left(1 - \dfrac{a^2}{s^2 + a^2} - 1 + \dfrac{b^2}{s^2 + b^2}\right)\right\}$$

$$= \dfrac{1}{b^2 - a^2}\, (b\, \sin\, bt - a\, \sin\, at)$$

1.16. Laplace's solution of linear differential equations with constant coefficients.

143

Example 1.

Solve the equation Y'''(t) - 6 Y''(t) +11Y'(t) — 6Y(t) = 1, given that Y(o) = Y'(0) = Y''(0) = 0.

Solution:

Transforming the equation i.e., multiplying both sides by e^{-st} and integrating w.r.t. t between o and ∞ p we get :

$$s^3 y(s) - 6 s^2 y(s) + 11 s y(s) - 6 y(s) = \frac{1}{s}$$

$$\therefore y(s) = \frac{1}{s(s^3 - 6 s^2 + 11 s - 6)}$$

$$= \frac{1}{s(s-1)(s-2)(s-3)}$$

$$= -\frac{1}{6s} + \frac{1}{2(s-1)} - \frac{1}{2(s-2)} + \frac{1}{6(s-3)}$$

Performing the inverse transformation we get

$$Y(t) = -\frac{1}{6} + \frac{1}{2} e^t - \frac{1}{2} e^{2t} + \frac{1}{6} e^{3t}$$

Example 2.

Solve the equation Y''(t) — 3 Y'(t) + 2 Y(t) = e^t, given that Y(0) = Y'(0) = 0

Solution:

$$(s^2 - 3s + 2) y(s) = \frac{1}{s-1}$$

$$\therefore y(s) = \frac{1}{(s-1)^2(s-2)}$$

$$= -\frac{1}{(s-1)^2} - \frac{1}{s-1} + \frac{1}{s-2}$$

$$\therefore Y(t) = -t e^t - e^t + e^{2t}$$

Example 3.

Solve the equation X''(t) + 2 X'(t) + X(t) = 3 t e^{-t}
given that X(0) = 4, X'(0) = 2

Solution:

144

$$s^2 x(s) - s\, X(o) - X'(o) + 2[s\, x(s) - X(o)] + x(s) = \frac{3}{(s+1)^2}$$

$$\therefore \quad s^2 x(s) - 4s - 2 + 2s\, x(s) - 8 + x(s) = \frac{3}{(s+1)^2}$$

$$(s^2 + 2s + 1)\, x(s) = 4s + 10 + \frac{3}{(s+1)^2}$$

$$\therefore \quad x(s) = \frac{4s}{(s+1)^2} + \frac{10}{(s+1)^3} + \frac{3}{(s+1)^4}$$

$$= \frac{4(s+1-1)}{(s+1)^2} + \frac{10}{(s+1)^3} + \frac{3}{(s+1)^4}$$

$$= \frac{4}{s+1} + \frac{6}{(s+1)^2} + \frac{3}{(s+1)^4}$$

$$\therefore \quad X(t) = 4e^{-t} + 6te^{-t} + 3e^{-t} \cdot \frac{t^3}{3!}$$

$$= e^{-t}\left(4 + 6t + \frac{1}{2}t^3\right)$$

Example 4.

Solve the equation $Y''(t) + 6\, Y'(t) + 9\, Y(t) = 6t^2 e^{-3t}$
given that $Y(0) = Y'(0) = 0$

Solution:

$$(s^2 + 6s + 9)\, y(s) = \frac{6 \times 2}{(s+3)^3}$$

$$\therefore \quad y(s) = \frac{12}{(s+3)^5}$$

$$\therefore \quad Y(t) = 12\, e^{-3t}\frac{t^4}{4!} = \tfrac{1}{2}\, e^{-3t}\, t^4$$

Example 5

Solve the equation $Y''(t) + 2\, Y'(t) + Y(t) = t$
given that $Y(0) = 3, \ Y(1) = -1$

Solution:

$$s^2 y(s) - s\, Y(0) - Y'(0) + 2\,[sy(s) - Y(0)] + y(s) = \frac{1}{s^2}$$

Put $Y'(0) = B$

$$\therefore\quad s^2 y(s) + 3s - B + 2sy(s) + 6 + y(s) = \frac{1}{s^2}$$

$$(s^2 + 2s + 1)\, y(s) + 3s + 6 - B = \frac{1}{s^2}$$

$$y(s) = \frac{-3s - 6 + B}{(s+1)^2} + \frac{1}{s^2 (s+1)^2}$$

$$= \frac{-3(s+1) - 5 + B}{(s+1)^2} + \frac{1}{s^2 (s+1)^2}$$

$$= -\frac{3}{s+1} + \frac{B-3}{(s+1)^2} + \frac{1}{s^2 (s+1)^2}$$

$$= -\frac{3}{s+1} + \frac{B-3}{(s+1)^2} - \frac{2}{s} + \frac{1}{s^2} + \frac{2}{s+1}$$

$$+ \frac{1}{(s+1)^2}$$

$$= -\frac{1}{s+1} + \frac{B-2}{(s+1)^2} - \frac{2}{s} + \frac{1}{s^2}$$

$$\therefore\quad Y(t) = -e^{-t} + (B-2)\, te^{-t} - 2 + t$$

Since $Y(1) = 1$, then

$$-1 = -e^{-1} + (B-2)\, e^{-1} - 2 + 1$$

$$= (B-3)\, e^{-1} - 1$$

$$\therefore\quad B - 3 = 0 \quad \text{i.e } B = 3$$

$$\therefore\quad Y(t) = (t-1)\, e^{-t} - 2 + t$$

Example 6
Solve the equation $Y'''(t) + Y(t) = 1$
given that $Y(0) = Y'(0) = Y''(0) = 0$

Solution:

146

$$(s^3 + 1) \, y(s) = \frac{1}{s}$$

$$\therefore \quad y(s) = \frac{1}{s(s^3+1)} = \frac{1}{s(s+1)(s^2-s+1)}$$

$$= \frac{1}{s} - \frac{1}{3(s+1)} - \frac{1}{3} \frac{2s-1}{s^2-s+1}$$

$$= \frac{1}{s} - \frac{1}{3(s+1)} - \frac{2}{3} \frac{s-\frac{1}{2}}{(s-\frac{1}{2})^2 + \frac{3}{4}}$$

$$\therefore \quad Y(t) = 1 - \frac{1}{3} e^{-t} - \frac{2}{3} e^{\frac{1}{2}t} \cos \frac{\sqrt{3}}{2} t$$

Example 7

Solve the equation $Y''(t) + k^2 \, Y(t) = 0$

Solution:

Put $Y(0) = A$, $Y'(0) = B$

$$s^2 y(s) - s Y(0) - Y'(0) + k^2 y(s) = 0$$

$$(s^2 + k^2) \, y(s) = As + B$$

$$y(s) = \frac{As}{s^2 + k^2} + \frac{B}{s^2 + k^2}$$

$$\left(\therefore \; Y(t) = A \cos kt + \frac{B}{k} \sin kt \right.$$

This is an example on simple harmonic oscillation, A and B being respectively the initial displacement and velocity.

Example 8

Solve the equation $X''(t) + 4 X(t) = 10 \sin 3t$
given that $X(0) = X'(0) = 0$

Solution:

$$(s^2+4) \, x(s) = 10 \times \frac{3}{s^2+9}$$

$$\therefore \quad x(s) = \frac{30}{(s^2+4)(s^2+9)}$$

$$= 6 \left(\frac{1}{s^2+4} - \frac{1}{s^2+9} \right)$$

$$\therefore \quad X(t) = 3 \sin 2t - 2 \sin 3t$$

This is an example on forced oscillation without damping.

Example 9
Solve the equation $Y''(t) + n^2 Y(t) = a \sin nt$
given that $Y(0) = Y'(0) = 0$

Solution:
$$(s^2 + n^2) y(s) = \frac{an}{s^2 + n^2}$$

$$\therefore y(s) = \frac{an}{(s^2 + n^2)^2}$$

Now $L^{-1} \left\{ \frac{1}{(s^2 + k^2)^2} \right\} = \frac{1}{2k^3} (\sin kt - kt \cos kt)$

$$\therefore Y(t) = \frac{an}{2n^3} (\sin nt - nt \cos nt)$$

$$= \frac{a}{2n^2} (\sin nt - nt \cos nt)$$

This is an example on resonance when the frequency of the impressed oscillations is equal to the natural frequency.

Example 10
Solve the equation $Y''(t) + 4Y'(t) + 13Y(t) = 5 \cos 3t$
given that $Y(0) = \frac{1}{4}$, $Y'(0) = 2$

Solution:

$$[s^2 y(s) - sY(o) - Y'(o)] + 4 [sy(s) - Y(o)] + 13y(s)$$

$$= \frac{5s}{s^2 + 9}$$

$$s^2 y(s) - \frac{1}{4} s - 2 + 4 [sy(s) - \frac{1}{4}] + 13y(s) = \frac{5s}{s^2 + 9}$$

$$(s^2 + 4s + 13) y(s) - \frac{1}{4} s - 3 = \frac{5s}{s^2 + 9}$$

$$y(s) = \frac{s + 12}{4 (s^2 + 4s + 13)} + \frac{5s}{(s^2 + 9) (s^2 + 4s + 13)}$$

$$= \frac{s^3 + 12s^2 + 2^?s + 108}{4\,(s^2 + 9)\,(s^2 + 4s + 13)}$$

$$= \frac{s + 9}{8(s^2 + 9)} + \frac{s + 11}{8(s^2 + 4s + 13)}$$

$$= \frac{1}{8}\,\frac{s}{s^2 + 9} + \frac{9}{8}\,\frac{1}{s^2 + 9} + \frac{1}{8}\,\frac{(s+2) + 9}{(s+2)^2 + 9}$$

$$= \frac{1}{8}\,\frac{s}{s^2 + 9} + \frac{9}{8}\,\frac{1}{s^2 + 9} + \frac{1}{8}\,\frac{s + 2}{(s+2)^2 + 9}$$

$$+ \frac{9}{8}\,\frac{1}{(s+2)^2 + 9}$$

$$\therefore Y(t) = \frac{1}{8}\cos 3t + \frac{3}{8}\sin 3t + \frac{1}{8}e^{-2t}\cos 3t + \frac{3}{8}e^{-2t}\sin 3t$$

$$= \frac{1}{8}\,(1 + e^{-2t})\,(\cos 3t + 3\sin 3t)$$

This is an example on forced oscillations with damping.

Example 11.
Solve the simultaneous differential equations
$$\dot{x} - x + 2y = 0$$
$$\dot{y} - 5x - 3y = 0$$
given that $x(0) = y(0) = 1$

Solution:

$$s\overline{x} - 1 - \overline{x} + 2\overline{y} = 0$$

$$s\overline{y} - 1 - 5\overline{x} - 3\overline{y} = 0$$

$$\therefore \quad (s - 1)\,\overline{x} + 2\overline{y} = 1$$

$$-5\overline{x} + (s-3)\,\overline{y} = 1$$

Eliminating $\overline{y}$ we get

$$[\,(s-1)\,(s-3) + 10\,]\,\overline{x} = s - 3 - 2$$

$$(s^2 - 4s + 13)\,\overline{x} = s - 5$$

$$\overline{x} = \frac{s - 5}{s^2 - 4s + 13} = \frac{(s-2) - 3}{(s-2)^2 + 9}$$

$$= \frac{s - 2}{(s-2)^2 + 9} - \frac{3}{(s-2)^2 + 9}$$

$$\therefore \quad x = e^{2t}\cos 3t - \frac{3}{3}e^{2t}\sin 3t$$

i.e. $\quad x = e^{2t}(\cos 3t - \sin 3t)$

Again eliminating $\overline{x}$ we get

$$[\,10 + (s-1)\,(s-3)\,]\,\overline{y} = 5 + s - 1$$

$$(s^2 - 4s + 13)\,\overline{y} = s + 4$$

$$\overline{y} = \frac{s + 4}{s^2 - 4s + 13} = \frac{s - 2 + 6}{(s-2)^2 + 9}$$

$$= \frac{s - 2}{(s-2)^2 + 9} + \frac{6}{(s-2)^2 + 9}$$

$$\therefore \quad y = e^{2t}(\cos 3t + 2\sin 3t)$$

1.1.7. Exercises on Laplace transformation

Find the Laplace transform of each of the following functions

(1) $3e^{5t}$ Ans. $\dfrac{3}{s-5}$, $s > 5$

(2) $4e^{-3t}$ Ans. $\dfrac{4}{s+3}$, $s > -3$

(3) $4t - 5$ Ans. $\dfrac{4}{s^2} - \dfrac{5}{s}$, $s > 0$

(4) $2\cos 6t$ Ans. $\dfrac{2s}{s^2 + 36}$, $s > 0$

(5) $6\sin 2t - 5\cos 2t$ Ans. $\dfrac{12-5s}{s^2+4}$, $s > 0$

(6) $t^2 - 3t + 5$ Ans. $\dfrac{2}{s^3} - \dfrac{3}{s^2} + \dfrac{5}{s}$, $s > 0$

(7) $\cos^2 kt$ Ans. $\dfrac{s^2 + 2k^2}{s(s^2 + 4k^2)}$, $s > 0$

(8) $(\sin t - \cos t)^2$ Ans. $\dfrac{s^2 - 2s + 4}{s(s^2 + 4)}$, $s > 0$

(9) $\cosh^2 4t$ Ans. $\dfrac{s^2 - 32}{s(s^2 - 64)}$, $s > |8|$

(10) $\sin t + 3\cos t$ Ans. $\dfrac{1 + 3s}{s^2 + 1}$, $s > 0$

(11) $F(t)$ $= 4$ $o < t < 1$ [Ans. $\dfrac{1}{s}(4 - e^{-s})$, $s > 0$]

  $= 3$ $t > 1$

(12) $\Phi(t) = \sin 2t$ $o < t < \pi$ [Ans. $\dfrac{2(1 - e^{-\pi s})}{s^2 + 4}$, $s > 0$]

  $= o$ $t > \pi$

(13) $t^{5/2}$ [Ans. $\dfrac{15}{8s^3}\left(\dfrac{\pi}{s}\right)^{1/2}$, $s > 0$]

(14) $t^3 e^{-2t}$ $\left[\text{Ans. } \dfrac{6}{(s+2)^4}\right]$

(15) $e^{-t} \cos 4t$ $\left[\text{Ans. } \dfrac{s+1}{s^2+2s+17}\right]$

(16) $e^{2t} \sin t$ $\left[\text{Ans. } \dfrac{1}{s^2-4s+5}\right]$

(17) $e^{-4t} \cosh 2t$ $\left[\text{Ans. } \dfrac{s+4}{s^2+8s+12}\right]$

Find the inverse transform of:

(18) $\dfrac{15}{s^2+4s+13}$ [Ans. $5e^{-2t} \sin 3t$]

(19) $\dfrac{s+1}{s^2+6s+25}$ [Ans. $e^{-3t} (\cos 4t - \tfrac{1}{2} \sin 4t)$]

(20) $\dfrac{s}{s^2-8s+16}$ [Ans. $e^{4t} (1+4t)$]

(21) $\dfrac{1}{s^2+2s+5}$ [Ans. $\tfrac{1}{2} e^{-t} \sin 2t$]

(22) $\dfrac{s}{s^2-6s+13}$ [Ans. $e^{3t} (\cos 2t + \dfrac{3}{2} \sin 2t)$]

(23) $\dfrac{1}{s^2+8s+16}$ [Ans. te^{-4t}]

(24) $\dfrac{s-5}{s^2+6s+13}$ [Ans. $e^{-3t}(\cos 2t - 4\sin 2t)$]

(25) $\dfrac{3s+1}{(s+1)^4}$ $\left[\text{Ans. } e^{-t}\left(\dfrac{3}{2}t^2 - \dfrac{1}{3}t^3\right)\right]$

(26) $\dfrac{s^2}{(s+2)^3}$ [Ans. $e^{-2t}(1-4t+2t^2)$]

(27) $\dfrac{s+2}{s^2-6s+8}$ [Ans. $3e^{4t} - 2e^{2t}$]

(28) $\dfrac{2s^2+5s-4}{s^3+s^2-2s}$ [Ans. $2+e^t-e^{-2t}$]

(29) $\dfrac{2s^2+1}{s(s+1)^2}$ [Ans. $1+e^{-t}-3t\,e^{-t}$]

(30) $\dfrac{1}{s^3(s^2+1)}$ [Ans. $\frac{1}{2}t^2-1+\cos t$]

(31) $\dfrac{5s-2}{s^2(s+2)(s-1)}$ [Ans. $t-2+e^t+e^{-2t}$]

(32) $\dfrac{s^3+s-4}{(s^2-2s+2)(s^2+2s-3)}$ [Ans. $e^t(\cos t + \sin t) - e^{-t}\sinh 2t$]

(33) $\dfrac{s}{(s^2+1)(s^2+3)}$ [Ans. $\frac{1}{2}(\cos t - \cos \sqrt{3}\, t)$]

(34) $\dfrac{5s+3}{(s-1)(s^2+2s+5)}$ [Ans. $e^t - e^{-t}(\cos 2t - \frac{3}{2}\sin 2t)$]

(35) $\dfrac{s+1}{(s^2+2s+2)^2}$ [Ans. $\frac{1}{2} t\, e^{-t}\sin t$]

(36) $\dfrac{s+1}{s(s^2+s-6)}$ [Ans. $-\dfrac{1}{6} + \dfrac{3}{10} e^{2t} - \dfrac{2}{15} e^{-3t}$]

Solve the following differential equations

(37) $Y'(t) - Y(t) = 4e^t \cos t$, given that $Y(0) = 3$

[Ans. $Y(t) = 3e^t + 4e^t \sin t$]

(38) $Y''(t) - 5Y'(t) + 6Y(t) = e^{2t}$, given that $Y(0)=1$, $Y'(0)=0$

[Ans. $Y(t) = (2-t) e^{2t} - e^{3t}$]

(39) $Y''(t) + 2Y'(t) + 5Y(t) = e^{-t}\sin t$, given that $Y(0) = 0$,

$Y'(0) = 1$ [Ans. $Y(t) = \frac{1}{3} e^{-t}(\sin t + \sin 2t)$]

(40) $Y^{(4)}(t) + 4Y'''(t) + 4Y''(t) = 0$, given that $Y(0) = 1$,

$Y'(0) = Y''(0) = Y'''(0) = 0$ [Ans. $Y(t) = 1$]

(41) $Y''(t) + Y(t) = t$ given that $Y(0) = 1$, $Y'(0) = -2$

[Ans. $Y(t) = t + \cos t - 3\sin t$]

(42) $Y''(t) - 3Y'(t) + 2Y(t) = 4e^{2t}$, given that $Y(0) = -3$,

$Y'(0) = 5$ [Ans. $Y(t) = -7e^t + 4e^{2t} + 4te^{2t}$]

(43) $Y''(t) + 9Y(t) = \cos 2t$, given that $Y(o) = 1$, $Y(\frac{\pi}{2})=-1$

[Ans. $Y(t) = \dfrac{4}{5}\cos 3t + \dfrac{4}{5}\sin 3t + \dfrac{1}{5}\cos 2t$]

(44) $Y''(t) + 2Y'(t) + Y(t) = \sin 2t$, given that $Y(0) = Y'(0)=0$

[Ans. $Y(t) = \dfrac{4}{25} e^{-t} + \dfrac{2}{5} te^{-t} - \dfrac{4}{25}\cos 2t - \dfrac{3}{25}\sin 2t$]

(45) $Y'(t) + Y(t) = t^2 e^{-t}$ given that $Y(0) = Y_o$

[Ans. $Y(t) = \frac{1}{3} t^3 e^{-t} + Y_o e^{-t}$]

(46) $x''(t) + 4x'(t) + 4x(t) = 4e^{-2t}$ given that $x(0) = -1$, $x'(0) = 4$ [Aus. $x(t) = e^{-2t}(2t^2 + 2t - 1)$]

(47) $x''(t) + x(t) = 6 \cos 2t$, given that $x(0) = 3$, $x'(0) = 1$

[Ans. $x(t) = 5 \cos t + \sin t - 2 \cos 2t$]

(48) $Y'''(t) - 3Y''(t) + 3Y'(t) - Y(t) = t^2 e^{t}$, given that $Y(0) = 1$, $Y'(0) = 0$, $Y''(0) = -2$

[Ans. $Y(t) = (1 - t - \frac{1}{2} t^2 + \frac{1}{60} t^5) e^{t}$]

(49) $y''(x) + 9y(x) = 40e^{x}$, $y(0) = 5$, $y'(0) = -2$

[Ans. $y(x) = 4e^{x} + \cos 3x - 2 \sin 3x$]

(50) $x''(t) - 4x'(t) + 4x(t) = e^{2t}$, $x'(0) = 0$, $x(1) = 0$

[Ans. $x(t) = \frac{1}{2} (1-t)^2 e^{2t}$]

(51) $x'''(t) - 3x''(t) + 3x'(t) - x(t) = 16e^{3t}$, given that $x(0) = 0$, $x'(0) = 4$, $x''(0) = 6$

[Ans. $x(t) = 2e^{3t} - (5t^2 + 2) e^{t}$]

Solve the following simultaneous equations

(52) $3X'(t) + 2X(t) - Y(t) = t$

$2Y'(t) - X(t) + Y(t) = 5e^{-t}$

given that $X(0) = Y(0) = 0$

[Ans. $X(t) = 6e^{-t/6} - (t+1) e^{-t} + t - 5$

$Y(t) = 9e^{-t/6} + (t-2) e^{-t} + t - 7$]

(53) $(D-2) x + 3y = 0$

$\qquad 2x + (D-1) y = 0$

$\qquad$ given that $x(0) = 8$, $y(0) = 3$

$\qquad$ [Ans. $x = 5e^{-t} + 3e^{4t}$, $y = 5 e^{-t} - 2e^{4t}$]

(54) $x''(t) - x(t) + 5y'(t) = t$

$\qquad y''(t) - 4y(t) - 2x'(t) = - 2$

$\qquad$ given that $x(0) = x'(0) = y(0) = y'(0) = 0$

$\qquad$ [Ans. $x(t) = - t + 5 \sin t - 2 \sin 2t$,

$\qquad\qquad y(t) = 1 - 2 \cos t + \cos 2t$]

(55) $5\dot{x} - 2\dot{y} + 4x - y = e^{-t}$

$\qquad \dot{x} + 8x - 3y = 5e^{-t}$

$\qquad$ given that $x = y = 0$ when $t = 0$

$\qquad$ [Ans. $x = 2e^{-t} + e^{t} - 3e^{-2t}$,

$\qquad\qquad y = 3e^{-t} + 3e^{t} - 6e^{-2t}$]

2. GENERAL THEOREMS ON THE LAPLACE TRANSFORMATION

2.1. The unit step function

This is a function which is equal to zero when t <a and is equal to unity when t >a, a being a positive constant. Let this function be denoted by U(t—a), then

$\qquad U(t-a) = 0 \qquad t < a$

$\qquad\qquad\quad = 1 \qquad t > a$

Let us obtain the Laplace transform of this function,

$$L\left\{U(t-a)\right\} = \int_0^a e^{-st}\, 0\; dt + \int_a^\infty e^{-st}\, 1\; dt = \frac{e^{-as}}{s}, \quad s>0.$$

If $a = 0$ we have the following special case

$L\left\{U(t)\right\} = 1/s$ where $U(t) = 0,\; t<0$ and $U(t) = 1\; t >0$

Any function $F(t)$ multiplied by $U(t-a)$ will have a value zero for $t <a$ and $F(t)$ for $t >a$

$$\text{i.e}\quad U(t-a)\, F(t) = 0 \qquad 0<t<a,$$

$$= F(t) \qquad t>a.$$

2.2. The second translation or shifting property

We shall now show that, if a is a positive constant and if

$$L^{-1}\left\{f(s)\right\} = F(t) \quad,\quad \text{then}$$

$$L^{-1}\left\{e^{-as} f(s)\right\} = 0, \qquad 0<t<a$$

$$= F(t-a), \quad t>a$$

This can also be written according to the in the form

$$L^{-1}\left\{e^{-as} f(s)\right\} = U(t-a)\, F(t-a)$$

For

$$e^{-as} f(s) = e^{-as} \int_0^\infty e^{-st}\, F(t)\; dt = \int_0^\infty e^{-s(t+a)}\, F(t)\; dt$$

Put $t + a = \tau$, then $dt = d\tau$

When $t = 0,\; \tau = a$ and when $t \longrightarrow \infty,\; \tau \longrightarrow \infty$

$$\therefore \int_0^\infty e^{-s(t+a)}\, F(t)\; dt = \int_a^\infty e^{-s\tau}\, F(\tau-a)\; d\tau$$

$$= \int_{0}^{a} e^{-st} 0 \ dt + \int_{a}^{\infty} e^{-st} F(t-a) \ dt$$

$$= \int_{0}^{\infty} e^{-st} U(t-a) \ F(t-a) \ dt = L \left\{ U(t-a) \ F(t-a) \right\}$$

Hence if $\quad L^{-1} \left\{ f(s) \right\} = F(t)$, then

$$L^{-1} \left\{ e^{-as} f(s) \right\} = U(t-a) F(t-a)$$

i.e $\quad L^{-1} \left\{ e^{-as} f(s) \right\} = 0 \ , \quad 0 < t < a$

$$= F(t-a) \ , \quad t > a$$

We can now give a simple geometrical interpretation to he function U (t - a) F(t - a). Let the definition of F(t) be extended such that F(t) is equal to zero for negative values of t. Then the graph of U(t —a) F(t--a) is actually the graph obtained by shifting the graph of F(t) to the right a distance a.

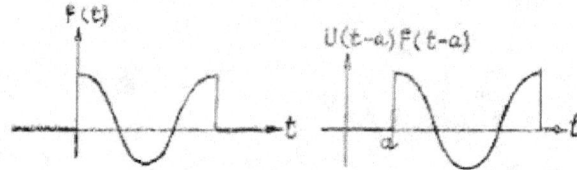

This can also be stated as follows:

If the definition of F(t) be extended such that F(t) is equal to zero for negative values of t, then the effect of shifting the graph of F(t) through a distance a to the right is to multiply the transform f(s) by e^{-as}.

Example 1.

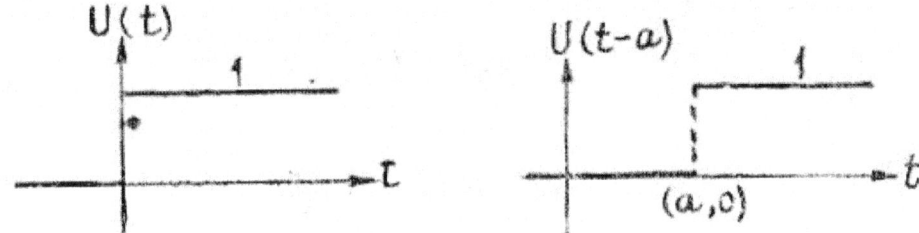

Since the unit step function U(t—a) is actually a shift of U(t) a distance a to the right and since L { U(t) } = 1 / s then

$$L\left\{ U_{(t-a)} \right\} = \frac{e^{-as}}{s}$$

Example 2

Since $L^{-1}\left\{ \frac{1}{s^4} \right\} = \frac{t^3}{3!}$, then

$$L^{-1}\left\{ \frac{e^{-as}}{s^4} \right\} = U(t-a)\frac{(t-a)^3}{3!}$$

i.e $L^{-1}\left\{ \frac{e^{-as}}{s^4} \right\} = 0 \qquad o < t < a$

$$= \frac{(t-a)^3}{3!} \qquad t > a$$

Example 3.

Since $L^{-1}\left\{ \frac{s}{s^2 + k^2} \right\} = \cos kt$, then

$$L^{-1}\left\{ \frac{e^{-as}\, s}{s^2 + k^2} \right\} = U(t-a)\cos k(t-a)$$

159

i.e $\quad L^{-1}\left\{\dfrac{e^{-as}\,s}{s^2+k^2}\right\} = 0 \qquad\qquad o < t < a$

$$= \cos k(t-a) \qquad t > a$$

Example 4

Find

$$L^{-1}\left\{\dfrac{(s+2)\,e^{-\pi s}}{s^2+s+1}\right\}$$

Solution:

$$L^{-1}\left\{\dfrac{s+2}{s^2+s+1}\right\} = L^{-1}\left\{\dfrac{(s+1/2)+3/2}{(s+1/2)^2+3/4}\right\}$$

$$= L^{-1}\left\{\dfrac{s+1/2}{(s+1/2)^2+3/4}\right\} + L^{-1}\left\{\dfrac{3/2}{(s+1/2)^2+3/4}\right\}$$

$$= e^{-\frac12 t}\cos\dfrac{\sqrt{3}}{2}t + \sqrt{3}\,e^{-\frac12 t}\sin\dfrac{\sqrt{3}}{2}t$$

$$= e^{-\frac12 t}\left(\cos\dfrac{\sqrt{3}}{2}t + \sqrt{3}\sin\dfrac{\sqrt{3}}{2}t\right)$$

$$\therefore\quad L^{-1}\left\{\dfrac{(s+2)\,e^{-\pi s}}{s^2+s+1}\right\} = e^{-\frac12(t-\pi)}\left[\cos\dfrac{\sqrt{3}}{2}(t-\pi)\right.$$

$$\left. + \sqrt{3}\sin\dfrac{\sqrt{3}}{2}(t-\pi)\right]U(t-\pi)$$

i.e $= 0 \qquad\qquad o < t < \pi$

$$= e^{-\frac12(t-\pi)}\left[\cos\dfrac{\sqrt{3}}{2}(t-\pi) + \sqrt{3}\sin\dfrac{\sqrt{3}}{2}(t-\pi)\right],\ t > \pi$$

Example 5

If $F(t) = L^{-1}\left\{\dfrac{e^{-3s}}{(s+1)^3}\right\}$, calculate $F(3/2)$, $F(4)$

$$L^{-1}\left\{\dfrac{1}{(s+1)^3}\right\} = \tfrac12\,t^2\,e^{-t}$$

$$\therefore \ F(t) = L^{-1} \left\{ \frac{e^{-3s}}{(s+1)^3} \right\} = \tfrac{1}{2}(t-3)^2 \ e^{-(t-3)} \ U(t-3)$$

i.e $F(t) = o \qquad o < t < 3$

$$F(t) = \tfrac{1}{2}(t-3)^2 \ e^{-(t-3)} \qquad , t > 3$$

$$\therefore \quad F(^3/_2) = o, \quad F(4) = \tfrac{1}{2}(4-3)^2 \ e^{-(4-3)} = \frac{1}{2e}$$

Example 6.

Find and sketch the function F(t) for which,

$$F(t) = L^{-1} \left\{ \frac{3}{s} - \frac{4e^{-s}}{s^2} + \frac{4e^{-3s}}{s^2} \right\}$$

Solution:

$$\text{Since } L^{-1} \left\{ \frac{1}{s^2} \right\} = t$$

$$\text{then} \quad L^{-1} \left\{ \frac{e^{-s}}{s^2} \right\} = (t-1) \ U(t-1)$$

$$\text{and} \quad L^{-1} \left\{ \frac{e^{-3s}}{s^2} \right\} = (t-3) \ U(t-3)$$

$$\therefore \quad F(t) = 3 - 4(t-1) \ U(t-1) + 4(t-3) \ U(t-3)$$

Hence for $\quad o < t < 1, \quad F(t) = 3$

for $\quad 1 < t < 3, \quad F(t) = 3 - 4(t-1) = 7 - 4t$

and for $\quad t > 3, \ F(t) = 3 - 4(t-1) + 4(t-3) = -5$

Hence the graph of F(t) is as shown in fig. 2-4

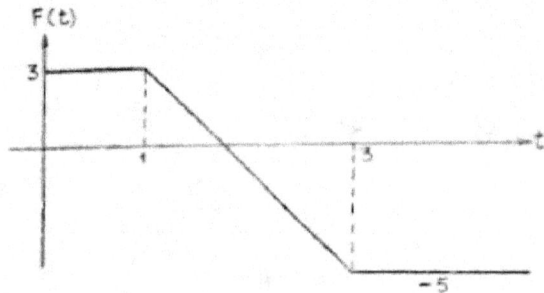

2.3. <u>Application</u> of the shift theorem to the solution of difference and differential equations

<u>Example</u> I

Solve the first order difference equation

$Y(t) = Y(t-h) + 1$ at $t \geqslant 0$

with the condition $Y(t) = 0$ at. $t < 0$
h, being a positive constant.

Solution:

The function $Y(t-h)$ is the same as the translated function $U(t-h) \, Y(t-h)$. Hence transforming the difference equation using the <u>shifting property</u> we get

$$y(s) = e^{-hs} y(s) + \frac{1}{s}$$

$$\text{i.e } \; y(s) = \frac{1}{s \, (1 - e^{-hs})}$$

$$= \frac{1}{s} \, (1 + e^{-hs} + e^{-2hs} + ..)$$

$$= \frac{1}{s} + \frac{1}{s} \, e^{-hs} + \frac{1}{s} \, e^{-2hs} + ..$$

$$\therefore \quad Y(t) = U(t) + U(t-h) + U(t-2h) + ..$$

$$\text{i.e } \; Y(t) = 1 \qquad\qquad o < t < h$$

$$= 1 + 1 = 2 \qquad h < t < 2h$$

$$= 1 + 1 + 1 = 3 \qquad 2h < t < 3h$$

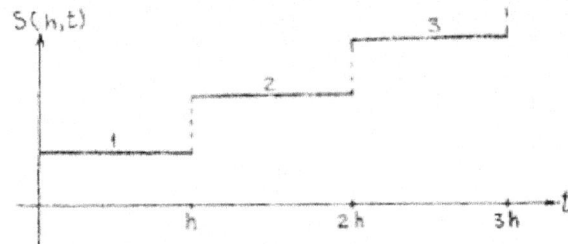

The function Y(t) is known as a <u>staircase function</u> S(h,t) with a positive run h and with a unit rise. Its transform is

$$\frac{1}{s(1-e^{-hs})} = \frac{1}{s} \cdot \frac{e^{\frac{hs}{2}}}{e^{\frac{hs}{2}} - e^{\frac{-hs}{2}}}$$

$$= \frac{1}{2s} \cdot \frac{\cosh\frac{hs}{2} + \sinh\frac{hs}{2}}{\sinh\frac{hs}{2}}$$

$$= \frac{1}{2s}\left(1 + \coth\frac{hs}{2}\right)$$

Example 2
Find the function Y(t) which satisfies the second order difference equation

$$Y(t) - 4Y(t - h) + 4Y(t - 2h) = 1$$

with the condition Y(t) = 0 when t < 0, h is a positive constant and the right hand side has to be replaced by zero when t < o.

Solution:

$$y(s) - 4e^{-hs} y(s) + 4e^{-2hs} y(s) = \frac{1}{s}$$

$$\therefore y(s)\left[1 - 2e^{-hs}\right]^2 = \frac{1}{s}$$

$$y(s) = \frac{1}{s(1-2e^{-hs})^2}$$

$$= \frac{1}{s}(1-2e^{-hs})^{-2}$$

$$= \frac{1}{s}\left[1 + 4e^{-hs} + \frac{-2\times-3}{2!}\times 4e^{-2hs} + .. \right]$$

$$= \frac{1}{s} + \frac{4}{s}e^{-hs} + \frac{12}{s}e^{-2hs} + ..$$

$\therefore$ $Y(t) = U(t) + 4U(t-h) + 12U(t-2h) + ..$

i.e $Y(t) = 1$ $o < t < h$

        $= 1+4=5$ $h < t < 2h$

        $= 1+4+12=17$ $2h < t < 3h$

Example 3
Solve the difference — differential equation
Y'(t) - 2Y(t-1) = 3
with the condition Y(t) = 0 when t ≤ 0 and the constant in the right hand side is to be replaced by zero when t < o.

Solution:

$$sy(s) - 2e^{-s}y(s) = \frac{3}{s}$$

$$(s-2e^{-s})y(s) = \frac{3}{s}$$

$$\therefore \; y(s) = \frac{3}{s(s-2e^{-s})} = \frac{3}{s^2\left(1-\dfrac{2e^{-s}}{s}\right)}$$

$$= \frac{3}{s^2}\left(1 + \frac{2e^{-s}}{s} + \frac{4e^{-2s}}{s^2} + ... \right)$$

$$= \frac{3}{s^2} + \frac{6e^{-s}}{s^3} + \frac{12e^{-2s}}{s^4} + ...$$

$$\therefore \; Y(t) = 3t + \frac{6}{2}(t-1)^2 U(t-1) + \frac{12}{3!}(t-2)^3 U(t-2) + ...$$

i.e $Y(t) = 3t$ $0 < t < 1$

$\qquad\qquad = 3t + 3(t-1)^2 \qquad\quad 1 < t < 2$

$\qquad\qquad = 3t + 3(t-1)^2 + 2(t-2)^3 \quad 2 < t < 3$

Example 4.

Compute

$$y\left(\frac{\pi}{2}\right) \quad \text{and} \quad y\left(2+\frac{\pi}{2}\right)$$

for the function $y(x)$ which satisfies the boundary value problem,

$$y''(x) + y(x) = (x-2)\ U(x-2) \quad y(o) = y'(o) = o$$

$$(x-2)\ U(x-2) = o \qquad\qquad 0 < x < 2$$

$$\qquad\qquad\qquad = x-2 \qquad\qquad x > 2$$

Transforming the differential equation we get

$$(s^2 + 1)\ \bar{y} = \frac{e^{-2s}}{s^2}$$

$$\therefore \quad \bar{y} = \frac{e^{-2s}}{s^2(s^2+1)} = e^{-2s}\left[\frac{1}{s^2} - \frac{1}{s^2+1}\right]$$

$$\therefore \quad y(x) = o \qquad\qquad 0 < x < 2$$

$$\qquad\qquad = x - 2 - \sin(x-2) \qquad x > 2$$

$$\therefore \quad y\left(\frac{\pi}{2}\right) = o$$

$$y\left(2+\frac{\pi}{2}\right) = 2 + \frac{\pi}{2} - 2 - \sin\left(2 + \frac{\pi}{2} - 2\right)$$

$$= \frac{\pi}{2} - 1$$

Example 5

Solve the equation

$$x''(t) + 4x(t) = \psi(t), \quad x(0) = 1, \quad x'(0) = 0 \text{ and where}$$
$\psi(t)$ is defined by

165

$$\psi(t) = 4t \qquad 0 \leqslant t \leqslant 1$$
$$= 4 \qquad\qquad t > 1$$

We write $\psi(t) = 4t - 4(t-1)\, U(t-1)$

$$\therefore \quad L\left\{ \psi(t) \right\} = \frac{4}{s^2} - \frac{4}{s^2}\, e^{-s}$$

$$L\left\{ x''(t) \right\} = s^2 \overline{x} - s$$

$$\therefore \quad s^2 \overline{x} - s + 4\overline{x} = \frac{4}{s^2} - \frac{4}{s^2}\, e^{-s}$$

$$(s^2 + 4)\, \overline{x} = s + \frac{4}{s^2} - \frac{4}{s^2}\, e^{-s}$$

$$\therefore \quad \overline{x} = \frac{s}{s^2 + 4} + \frac{4}{s^2(s^2 + 4)} - \frac{4e^{-s}}{s^2(s^2 + 4)}$$

$$= \frac{s}{s^2 + 4} + \frac{1}{s^2} - \frac{1}{s^2 + 4} - \left(\frac{1}{s^2} - \frac{1}{s^2 + 4} \right) e^{-s}$$

$$\therefore \quad x(t) = \cos 2t + t - \tfrac{1}{2} \sin 2t - [t - 1$$
$$- \tfrac{1}{2} \sin 2(t-1)]\, U(t-1)$$

2.4. The unit impulse function

Consider the function $F(t)$ defined by
$$F(t) = \frac{1}{\epsilon} \qquad 0 < t < \epsilon$$
$$= 0 \qquad\qquad t > \epsilon$$
The area under the function is unity

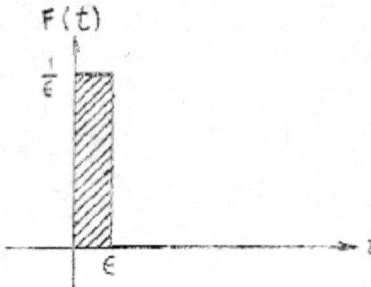

Its Laplace transform is

$$\int_0^\epsilon e^{-st} \cdot \frac{1}{\epsilon} \, dt = \frac{1}{\epsilon} \left[\frac{e^{-st}}{-s} \right]_0^\epsilon = \frac{1}{\epsilon s} (1 - e^{-\epsilon s})$$

Now let us take the limiting value of the transform when $\epsilon \longrightarrow 0$

$$\lim_{\epsilon \to 0} \frac{1 - e^{-\epsilon s}}{\epsilon s} = \lim_{\epsilon \to 0} \frac{s e^{-\epsilon s}}{s} = 1$$

The limiting function of F(t) when $\epsilon \longrightarrow 0$ is known as the unit impulse function or the Dirac delta function $\delta(t)$. Thus

$$L \{ \delta(t) \} = 1$$

If the function be shifted a distance T along the positive t axis, then according to the second shift theorem its transform is

$$L \{ \delta(t - T) \} = e^{-Ts}$$

we shall now show that the derivative of the unit step function is the impulse function. For consider the function

$$\frac{1}{\epsilon} U(t) - \frac{1}{\epsilon} U(t - \epsilon) \quad \ldots \quad (1)$$

Its transform is

$$\frac{1}{\epsilon} \left(\frac{1}{s} - \frac{1}{s} e^{-\epsilon s} \right) = \frac{1}{\epsilon s} \left(1 - e^{-\epsilon s} \right) \quad \ldots \quad (2)$$

and as $\epsilon \to 0$, the expression (1) tends formally to $U'(t)$ and its transform (2) tends to 1.

$$\text{Thus} \quad L \{ U'(t) \} = L \{ \delta(t) \} = 1$$

$$\text{and} \quad L \{ U'(t - T) \} = L \{ \delta(t - T) \} = e^{-Ts}$$

Example
A series circuit of resistance R, inductance. L is connected to a generator which delivers an impulse voltage $V_0 \delta(t)$ which is impressed upon the circuit at zero time. Find the current in the circuit.

Solution:

$$L \frac{dI}{dt} + RI = V_0 \, \delta(t)$$

$$\text{Lsi(s)} + \text{Ri(s)} = V_0, \quad I(0) = 0$$

$$(Ls+R)i(s) = V_0 \quad \therefore \quad i(s) = \frac{V_0}{L} \cdot \frac{1}{s+\dfrac{R}{L}}$$

$$\therefore \quad I(t) = \frac{V_0}{L} e^{-\frac{R}{L}t}$$

Note: An impulsive voltage means a large voltage acting for a very short interval of time but the time integral is finite.

2.5. The unit doublet

Consider the function F(t) defined by

$$F(t) = \frac{1}{\epsilon^2} \qquad 0 < t < \epsilon$$

$$= -\frac{1}{\epsilon^2} \qquad \epsilon < t < 2\epsilon$$

$$= 0 \qquad t > 2\epsilon$$

$$L\left\{F(t)\right\} = \int_0^\epsilon \frac{1}{\epsilon^2} e^{-st} \, dt + \int_\epsilon^{2\epsilon} -\frac{1}{\epsilon^2} e^{-st} \, dt$$

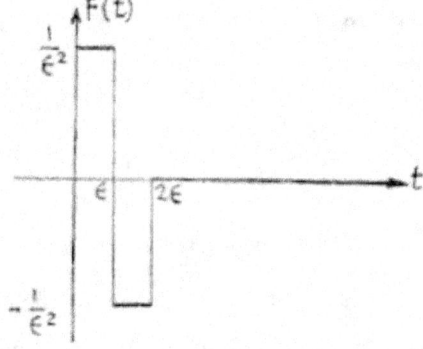

$$= \frac{1}{\epsilon^2 s}\left[1 - 2e^{-\epsilon s} + e^{-2\epsilon s} \right]$$

$$= \frac{1}{\epsilon^2 s}\left(1 - e^{-\epsilon s} \right)^2$$

Let us find the limiting value of this transform when $\epsilon \rightarrow 0$

$$\lim_{\epsilon \rightarrow 0} \frac{1 - 2e^{-\epsilon s} + e^{-2\epsilon s}}{\epsilon^2 s} = \lim_{\epsilon \rightarrow 0} \frac{-2s^2 e^{-\epsilon s} + 4s^2 e^{-2\epsilon s}}{2s}$$

$$= \frac{2s^2}{2s} = s$$

The limiting function of $F(t)$ when $\epsilon \rightarrow 0$ is known as the unit doublet and its transform is s. If the function is shifted along the positive t axis a distance T then its transform is $s\,e^{-Ts}$.

We shall now show that the derivative of the unit impulse function is the unit doublet. For consider the function

$$\frac{1}{\epsilon} \delta(t) - \frac{1}{\epsilon} \delta(t - \epsilon) \quad . \quad . \quad . \quad (1)$$

whore ϵ may he made as small as we please. The Laplace transform of this is

$$\frac{1}{\epsilon} (1 - e^{-\epsilon s}) \quad . \quad . \quad . \quad . \quad (2)$$

and as $\epsilon \rightarrow 0$ the expression (1) tends formally to $\delta'(t)$ and its transform (2) tends to s

Thus $L\left\{ \delta'(t) \right\} = s$

and $L\left\{ \delta'(t - T) \right\} = se^{-Ts}$

In other words

$$L\left\{ U''(t) \right\} = s$$

and $L\left\{ U''(t - T) \right\} = s\,e^{-Ts}.$

2.6. The behavior of f(s) as $s \rightarrow \infty$

Let F(t) be a function which is sectionally continuous in every finite interval $0 \leq t \leq T$ and let it be of the order of $e^{a_0 t}$ as $t \to \infty$, i.e., $|F(t)| < Me^{a_0 t}$.

Take $s \geq a$ where $a \geq a_0$ then

$$\left| e^{-st} F(t) \right| < e^{-at} Me^{a_0 t} = Me^{-(a-a_0)t}$$

$$\therefore \quad \left| f(s) \right| = \left| L\left\{ F(t) \right\} \right| = \left| \int_0^{\infty} e^{-st} F(t)\, dt \right|$$

$$\leq \int_0^{\infty} \left| e^{-st} F(t) \right| dt$$

$$< \int_0^{\infty} Me^{-(a-a_0)t} dt$$

$$= M\left[\frac{e^{-(a-a_0)t}}{-(a-a_0)} \right]_0^{\infty}$$

$$= \frac{M}{a-a_0} \quad , \quad s \geq a$$

Put $a = s$ $\therefore |f(s)| < \dfrac{M}{s-a_0} \longrightarrow 0$ as $s \longrightarrow \infty$

$$\therefore f(s) \longrightarrow 0 \text{ as } s \longrightarrow \infty$$

2.7. Initial value theorem

To prove that $\displaystyle\lim_{t \to 0} F(t) = \lim_{s \to \infty} s f(s)$

$$L\left\{ F'(t) \right\} = \int_0^{\infty} e^{-st} F'(t)\, dt = s f(s) - F(o) \ldots \qquad (1)$$

Now if F'(t) is sectionally continuous and of exponential order then according to the theorem in 2.6.

170

$$\lim_{s \to \infty} \int_{0}^{\infty} e^{-st} F'(t) \, dt = 0$$

Taking the limit in (1), assuming $F(t)$ to be continuous at $t = 0$, we get

$$0 = \lim_{s \to \infty} s f(s) - F(o)$$

i.e $\lim_{s \to \infty} s f(s) = F(o) = \lim_{t \to 0} F(t).$

2.8. Final value theorem

To prove that $\lim_{t \to \infty} F(t) = \lim_{s \to 0} s f(s)$

$$L \left\{ F'(t) \right\} = \int_{0}^{\infty} e^{-st} F'(t) \, dt = s f(s) - F(o) \quad . \quad . \quad . \quad (1)$$

The limit of the left hand side of (1) as $s \to 0$ is

$$\lim_{s \to 0} \int_{0}^{\infty} e^{-st} F'(t) dt = \int_{0}^{\infty} F'(t) dt = [F(t)]_{0}^{\infty}$$

$$= \lim_{t \to \infty} F(t) - F(o)$$

The limit of right hand side of (1) as $s \to 0$ is

$$\lim_{s \to 0} s f(s) - F(o)$$

$$\therefore \lim_{t \to \infty} F(t) - F(o) = \lim_{s \to 0} s f(s) - F(o)$$

i.e $\lim_{t \to \infty} F(t) = \lim_{s \to 0} s f(s)$

Example

Let $F(t) = 2 e^{-3t}$, then $f(s) = \dfrac{2}{s+3}$

By the initial value theorem we have

$$\lim_{t \to 0} 2e^{-3t} = \lim_{s \to \infty} \frac{2s}{s+3} = 2$$

and by the final value theorem we have

$$\lim_{t \to \infty} 2e^{-3t} = \lim_{s \to 0} \frac{2s}{s+3} = 0$$

2.9. Differentiation of transform

Let F(t) be a sectionally continuous function in every finite interval $o \le t \le T$ and let it be of the order of e^{at} as $t \to \infty$. Then if s>a, we have

$$L\left\{F(t)\right\} = \int_0^\infty e^{-st} F(t)\, dt = f(s)$$

Differentiate w.r.t. s

$$\frac{\partial}{\partial s} \int_0^\infty e^{-st} F(t)\, dt = \int_0^\infty \frac{\partial}{\partial s}\left\{ e^{-st} F(t) \right\} dt$$

$$= \int_0^\infty - te^{-st} F(t)\, dt$$

$$= L\left\{ - t\, F(t) \right\} = f'(s)$$

Similarly $\quad L\left\{(-t)^2 F(t) \right\} = f''(s)$

and in general $\quad L\left\{ (-t)^n F(t)\right\} = f^{(n)}(s)$

Hence the differentiation of the transform of a function w. r. t. s corresponds to the multiplication of the function by - t.

e.g. $\quad L\left\{ \sin kt \right\} = \dfrac{k}{s^2+k^2}, \quad s > 0$

$\therefore \quad L\left\{ -t \sin kt \right\} = \dfrac{-2ks}{(s^2 + k^2)^2}$

Then $L^{-1}\left\{\dfrac{s}{(s^2+k^2)^2}\right\} = \dfrac{1}{2k}\,t\,\sin kt$

Similarly $L\left\{\cos kt\right\} = \dfrac{s}{s^2+k^2}$

$\therefore L\left\{-t\cos kt\right\} = \dfrac{s^2+k^2-2s^2}{(s+k^2)^2}$

$\therefore L\left\{t\cos kt\right\} = \dfrac{s^2-k^2}{(s^2+k^2)^2} = \dfrac{s^2+k^2-2k^2}{(s^2+k^2)^2}$

$\qquad = \dfrac{1}{s^2+k^2} - \dfrac{2k^2}{(s^2+k^2)^2}$

$\qquad = L\left\{\dfrac{\sin kt}{k}\right\} - \dfrac{2k^2}{(s^2+k^2)^2}$

$\therefore L\left\{t\cos kt - \dfrac{1}{k}\sin kt\right\} = -\dfrac{2k^2}{(s^2+k^2)^2}$

i.e. $L^{-1}\left\{\dfrac{1}{(s^2+k^2)^2}\right\} = \dfrac{1}{2k^3}(\sin kt - kt\cos kt)$

2.10. <u>Application</u> of the differentiation of Laplace transform to the solution of linear differential equations with coefficients as polynomials in t.

<u>Example</u> 1
Find a solution of the differential equation

$$t\frac{d^2x}{dt^2} + t\frac{dx}{dt} + x = 0$$

which satisfies $x = 0$, $dx/dt = 1$ when $t = 0$.

$$-\frac{d}{ds}\left[s^2\bar{x} - sx(0)-x'(0)\right] - \frac{d}{ds}\left[s\bar{x} - x(0)\right] + \bar{x} = 0$$

$$\therefore -\frac{d}{ds}\left[s^2\bar{x} - 1\right] - \frac{d}{ds}s\bar{x} + \bar{x} = 0$$

$$-s^2\frac{d\bar{x}}{ds} - 2s\bar{x} - s\frac{d\bar{x}}{ds} - \bar{x} + \bar{x} = 0$$

173

$$\therefore \quad (1+s)\frac{d\bar{x}}{ds} + 2\ \bar{x} = 0$$

$$\int \frac{d\bar{x}}{\bar{x}} + 2 \int \frac{ds}{1+s} = 0$$

$$\log\ \bar{x} + 2 \log (1+s) = \log C$$

$$\therefore \quad \bar{x} = \frac{C}{(1+s)^2}$$

$$x = cte^{-t}$$

The condition x =1 when t = 0 is satisfied and the condition dx/dt = 1 when t = 0 gives
C = 1
Hence

$$x = te^{-t}$$

Example 2

Find the solution of Bessel's differential equation of zero order

t Y"(t) + Y'(t) t Y(t) = 0

which has a Laplace transform and which satisfies Y(0)=1

Solution:

$$-\frac{d}{ds}\ [s^2 y(s) - s - Y'(0)] + [\ sy(s) - 1\ ] - \frac{d}{ds}\ y(s) = 0$$

$$\therefore \quad -2sy - s^2\frac{dy}{ds} + 1 + sy - 1 - \frac{dy}{ds} = 0$$

$$\text{giving}\ (s^2 + 1)\frac{dy}{ds} + sy = 0$$

$$\therefore \quad \frac{dy}{y} = \frac{-sds}{s^2 + 1}$$

$$\log y = -\tfrac{1}{2}\log (s^2+1) + \log C$$

$$\therefore \quad y(s) = \frac{C}{\sqrt{s^2 + 1}}$$

Now by the initial value theorem

$$\lim_{t \to 0} Y(t) = \lim_{s \to \infty}\ sy(s)$$

$$\therefore \; 1 = \lim_{s \to \infty} \frac{Cs}{\sqrt{s^2+1}} = C$$

also C can be determined as follows:

$$y(s) = \frac{C}{s}\left(1 + \frac{1}{s^2}\right)^{-\frac{1}{2}}$$

$$= \frac{C}{s}\left(1 - \frac{1}{2} \cdot \frac{1}{s^2} - \frac{1}{2} \times -\frac{3}{2} \times \frac{1}{2!} \cdot \frac{1}{s^4} \cdots \right)$$

$$= \frac{C}{s}\left(1 - \frac{1}{2}\frac{1}{s^2} + \frac{1.3}{2^2.2!} \cdot \frac{1}{s^4} \cdots \right)$$

$$= C\left(\frac{1}{s} - \frac{1}{2} \cdot \frac{1}{s^3} + \frac{1.3}{2^2.2!} \cdot \frac{1}{s^5} - \cdots \right)$$

$$\therefore \; Y(t) = C\left[1 - \frac{1}{2} \cdot \frac{t^2}{2!} + \frac{1.3}{2^2.2!} \cdot \frac{t^4}{4!} \cdots \right]$$

and since $Y(o) = 1$ $\therefore$ C=1 and we get Bessel function of zero order of the first kind namely

$$J_0(t) = 1 - \frac{t^2}{2^2} + \frac{t^4}{2^2.4^2} - \frac{t^6}{2^2.4^2.6^2} + \cdots$$

Corollary 1.

$$L\left\{J_0(t)\right\} = \frac{1}{\sqrt{s^2+1}} \; , \quad s > o$$

Corollary 2.

$$L\left\{J_0(at)\right\} = \frac{1}{a} \cdot \frac{1}{\sqrt{\left(\frac{s}{a}\right)^2+1}} = \frac{1}{\sqrt{s^2+a^2}}$$

Corollary 3.

Since $J_0'(t) = - J_1(t)$

then $L\left\{J_0'(t)\right\} = L\left\{- J_1(t)\right\}$

$$s\, L\left\{J_0(t)\right\} - J_0(o) = L\left\{- J_1(t)\right\}$$

$$\frac{s}{\sqrt{s^2+1}} - 1 = - L\left\{J_1(t)\right\}$$

$$\therefore \ L \left\{ J_1(t) \right\} = 1 - \frac{s}{\sqrt{s^2+1}} = \frac{\sqrt{s^2+1}-s}{\sqrt{s^2+1}}$$

$$= \frac{1}{\sqrt{s^2+1} \ (s+\sqrt{s^2+1} \ )}$$

Example 3

Find the Laplace transform of $\sin\sqrt{t}$

Solution:

Put $Y(t) = \sin\sqrt{t}$

By differentiating twice we get

$$4t \ Y''(t) + 2 \ Y'(t) + Y(t) = 0$$

Transforming we get

$$- 4 \frac{d}{ds}[s^2y - sY(o) - Y'(o)] + 2[sy - Y(o)] + y = o$$

$$\therefore \ 4s^2\frac{dy}{ds} + (6s-1)y = o$$

$$y = \frac{C}{s^{3/2}} \ e^{-\frac{1}{4s}}$$

For Small values of t, $\sin\sqrt{t} = \sqrt{t}$ approx.

$$\text{and } L \left\{ \sqrt{t} \right\} = \frac{\sqrt{\pi}}{2s^{3/2}} \qquad \text{✳}$$

For large s, $y = \frac{C}{s^{3/2}}$ approx Hence by comparison $C = \frac{\sqrt{\pi}}{2}$

Alternative method

$$\sin\sqrt{t} = \sqrt{t} - \frac{(\sqrt{t})^3}{3!} + \frac{(\sqrt{t})^5}{5!} - \cdots$$

$$= t^{1/2} - \frac{t^{3/2}}{3!} + \frac{t^{5/2}}{5!} - \cdots$$

176

$$\therefore \quad L\left\{\sin \sqrt{t}\right\} = \frac{\Gamma(^3/_2)}{s^{3/_2}} - \frac{\Gamma(^5/_2)}{3! \; s^{5/_2}} + \frac{\Gamma(^7/_2)}{5! \; s^{7/_2}} - \cdots$$

$$= \frac{\sqrt{\pi}}{2s^{3/_2}}\left[1 - (1/2^2 s) + \frac{(1/2^2 s)^2}{2!} - \cdots\right]$$

$$= \frac{\sqrt{\pi}}{2s^{3/2}} \; 0^{-\frac{1}{2^2 s}} = \frac{\sqrt{\pi}}{2s^{3/_2}} \; e^{-\frac{1}{4s}}$$

2.11. Integration of transforms

Let F(t) be sectionally continuous of the order of e^{at} as $t \longrightarrow \infty$ and let x > a, then

$$f(x) = \int_0^\infty e^{-xt} F(t) \; dt$$

Now let s > a and suppose that $\lim\limits_{t \longrightarrow +0} \frac{F(t)}{t}$ exists, then

$$\int_s^\infty f(x) \; dx = \int_s^\infty \left\{\int_0^\infty e^{-xt} F(t) \; dt\right\} dx$$

Assuming that we can invert the order of integration which is true in this case, we have

$$\int_s^\infty f(x) \; dx = \int_0^\infty F(t) \left\{\int_s^\infty e^{-xt} \; dx\right\} dt$$

$$= \int_0^\infty F(t) \left[\frac{e^{-xt}}{-t}\right]_s^\infty dt = \int_0^\infty \frac{F(t)}{t} e^{-st} \; dt$$

$$= \int_0^\infty e^{-st} \left\{\frac{F(t)}{t}\right\} dt = L\left\{\frac{F(t)}{t}\right\}$$

$$\therefore \quad \text{If} \quad L\left\{F(t)\right\} = f(s)$$

$$\text{then} \quad L\left\{\frac{F(t)}{t}\right\} = \int_s^\infty f(x) \; dx$$

Hence the division of the object function by t corresponds to the integration of the transform of that function from s to ∞.

177

Example 1.

Prove that

$$L\left\{\frac{1-\cos kt}{t}\right\} = \tfrac{1}{2}\log\left(1+\frac{k^2}{s^2}\right), \quad s > k > 0.$$

Solution:

$$L\left\{1-\cos kt\right\} = \frac{1}{s} - \frac{s}{s^2+k^2}$$

$$\therefore L\left\{\frac{1-\cos kt}{t}\right\} = \int_s^\infty \left(\frac{1}{x} - \frac{x}{x^2+k^2}\right) dx$$

$$= \left[\log x - \tfrac{1}{2}\log(x^2+k^2)\right]_s^\infty$$

$$= \lim_{x\to\infty}\log\frac{x}{\sqrt{x^2+k^2}} - \left[\log\frac{x}{\sqrt{x^2+k^2}}\right]_{x=s}$$

$$= 0 - \log\frac{s}{\sqrt{s^2+k^2}}$$

$$= \tfrac{1}{2}\log\frac{s^2+k^2}{s^2} = \tfrac{1}{2}\log\left(1+\frac{k^2}{s^2}\right)$$

Example 2.

$$L\left\{e^{-at} - e^{-bt}\right\} = \frac{1}{s+a} - \frac{1}{s+b}$$

$$\therefore L\left\{\frac{e^{-at} - e^{-bt}}{t}\right\} = \int_s^\infty \left[\frac{1}{x+a} - \frac{1}{x+b}\right] dx$$

$$= \left[\log\frac{x+a}{x+b}\right]_s^\infty = -\log\frac{s+a}{s+b} = \log\frac{s+b}{s+a}$$

$$\therefore L\left\{\frac{e^{-at} - e^{-bt}}{t}\right\} = \log\frac{s+b}{s+a}, \quad s > -a, s > -b$$

Corollary

If a = 0 and b = 1 we get

$$L\left\{\frac{1-e^{-t}}{t}\right\} = \log\frac{s+1}{s}$$

$$\therefore L\left\{\frac{1-e^{-t}}{t}\right\} = \log\left(1+\frac{1}{s}\right), s > 0$$

Example 3

Prove that $\displaystyle\int_0^\infty \frac{F(t)}{t}\,dt = \int_0^\infty f(x)\,dx$ provided that the integrals converge and hence deduce that

(i) $\displaystyle\int_0^\infty \frac{\sin t}{t}\,dt = \frac{\pi}{2}$, (ii) $\displaystyle\int_0^\infty \frac{e^{-t}-e^{-3t}}{t}\,dt = \log 3$

Solution:

Since $L\left\{\dfrac{F(t)}{t}\right\} = \displaystyle\int_s^\infty f(x)\,dx$, then

$$\int_0^\infty e^{-st}\frac{F(t)}{t}\,dt = \int_s^\infty f(x)\,dx$$

Let s → 0 + and assuming that the integrals converge, then

$$\int_0^\infty \frac{F(t)}{t}\,dt = \int_0^\infty f(x)\,dx$$

(i) Take $F(t) = \sin t$, then $f(s) = \dfrac{1}{s^2+1}$ and we get

$$\int_0^\infty \frac{\sin t}{t}\,dt = \int_0^\infty \frac{dx}{x^2+1} = \left[\tan^{-1}x\right]_0^\infty = \frac{\pi}{2}$$

179

(ii) Take $F(t) = e^{-t} - e^{-3t}$, then $f(s) = \dfrac{1}{s+1} - \dfrac{1}{s+3}$

$$\therefore L\left\{\frac{e^{-t} - e^{-3t}}{t}\right\} = \int_s^\infty \left[\frac{1}{x+1} - \frac{1}{x+3}\right] dx$$

$$\therefore \int_o^\infty e^{-st}\left(\frac{e^{-t} - e^{-3t}}{t}\right) dt = \left[\log\frac{x+1}{x+3}\right]_s^\infty = \log\frac{s+3}{s+1}$$

Taking the limit as s → 0 we get

$$\int_o^\infty \frac{e^{-t} - e^{-3t}}{t}\, dt = \log 3$$

2.12. Transforms of periodic functions

If F(t) is a periodic function of period a, i.e., F(t +a) = F(t) and if F(t) is sectionally continuous over a period, then

$$f(s) = \frac{\displaystyle\int_0^a e^{-st}\, F(t)\, dt}{1 - e^{-as}}$$

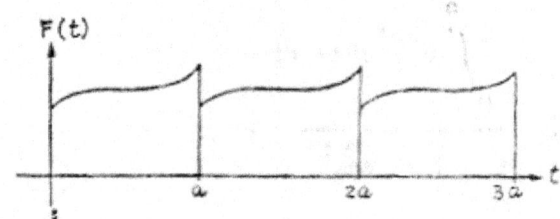

$$f(s) = \int_0^\infty e^{-st} F(t)\, dt = \int_0^a e^{-st} F(t)\, dt + \int_a^{2a} e^{-st} F(t)\, dt + \ldots$$

$$= \sum_{n=0}^\infty \int_{na}^{(n+1)a} e^{-st} F(t)\, dt$$

Put $t = \tau + na$, then $F(\tau + na) = F(\tau)$

$$f(s) = \sum_{n=0}^\infty \int_0^a e^{-s(\tau+na)} F(\tau+na)\, d\tau$$

$$= \sum_{n=0}^\infty e^{-nas} \int_0^a e^{-s\tau} F(\tau)\, d\tau$$

$$= \sum_{n=0}^\infty e^{-nas} \int_0^a e^{-st} F(t)\, dt$$

$$= (1 + e^{-as} + e^{-2as} + e^{-2as} + \ldots) \int_0^a e^{-st} F(t)\, dt$$

$$= \frac{\displaystyle\int_0^a e^{-st} F(t)\, dt}{1 - e^{-as}}$$

Example 1

Find the Laplace transform of the square wave or the Meander function M(c,t) defined by

$$M(c,t) = 1 \qquad 0 < t < c$$
$$= -1 \qquad c < t < 2c$$
$$M(c,t+2c) = M(c,t)$$

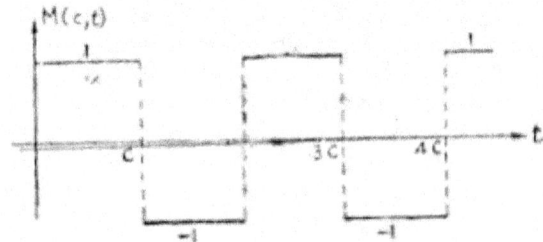

Since the period is 2c, then

$$L\left\{M(c,t)\right\} = \frac{\displaystyle\int_0^{2c} e^{-st}\ M(c,t)\ dt}{1-e^{-2cs}}$$

Now $\displaystyle\int_0^{2c} e^{-st}\ M(c,t)\ dt = \int_0^{c} e^{-st}\ dt - \int_c^{2c} e^{-st}\ dt$

$$= \left[\frac{e^{-st}}{-s}\right]_0^{c} + \left[\frac{e^{-st}}{s}\right]_c^{2c} = \frac{1}{s}\left(1-e^{-cs}\right)^2$$

$$\therefore\ L\left\{M(c,t)\right\} = \frac{1}{s}\ \frac{(1-e^{-cs})^2}{1-e^{-2cs}} = \frac{1}{s}\ \frac{1-e^{-cs}}{1+e^{-cs}}$$

$$= \frac{1}{s}\ \frac{e^{\frac{cs}{2}}-e^{-\frac{cs}{2}}}{e^{\frac{cs}{2}}+e^{-\frac{cs}{2}}}$$

$$= \frac{1}{s}\ \tanh\ \frac{cs}{2}\ ,\ s>0$$

Example 2.
Find the Laplace transform of the triangular wave or the function H(c,t) defined as follows :

182

$$H(c,t) = t \qquad 0 < t < c$$
$$= 2c-t \qquad c < t < 2c$$
$$H(c,t+2c) = H(c,t)$$

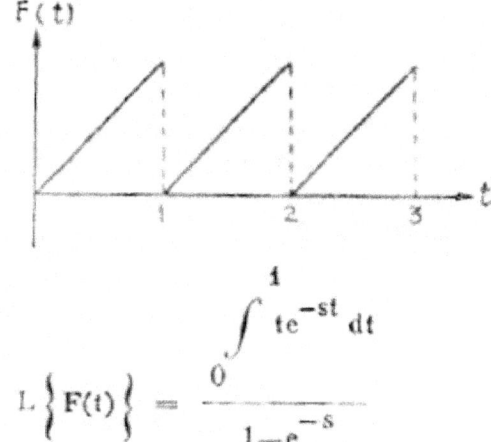

It is evident that H(c,t) at any t is the integral of the function M(c,t) from o to t.

$$\therefore \quad L\left\{ H(c,t) \right\} = \frac{1}{s^2} \; \tanh \frac{cs}{2}$$

Example 3
Find the Laplace transform of the periodic function F(t) defined by

$$F(t) = t \qquad 0 < t < 1$$
$$F(t+1) = F(t)$$

$$L\left\{ F(t) \right\} = \frac{\displaystyle\int_0^1 te^{-st} \, dt}{1-e^{-s}}$$

$$\int_0^1 te^{-st}\, dt = -\frac{1}{s}\int_0^1 t d e^{-st} = -\frac{1}{s}\left[\ te^{-st} + \frac{e^{-st}}{s}\ \right]_0^1$$

$$= -\frac{1}{s}\left[\ e^{-s} + \frac{e^{-s}}{s} - \frac{1}{s}\ \right] = \frac{1}{s^2} - \frac{s+1}{s^2}e^{-s}$$

$$= \frac{1}{s^2}\left(1 - e^{-s}\right) - \frac{1}{s}e^{-s}$$

$$\therefore\ L\left\{\ F(t)\ \right\} = \frac{1}{s^2} - \frac{e^{-s}}{s(1-e^{-s})}$$

Example 4

Find the Laplace transform of the intermittent sine wave defined by

$$F(t) = \sin t \qquad 0 < t < \pi$$

$$= 0 \qquad \pi < t < 2\pi$$

$$F(t+2\pi) = F(t)$$

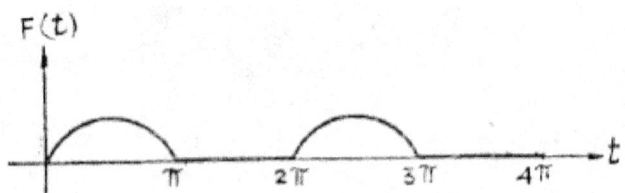

$$L\left\{\ F(t)\ \right\} = \frac{\displaystyle\int_0^{2\pi} e^{-st} F(t)\, dt}{1 - e^{-2\pi s}} = \frac{\displaystyle\int_0^{\pi} e^{-st}\sin t\, dt}{1 - e^{-2\pi s}}$$

$$\text{Now } \int_0^{\pi} e^{-st}\sin t\, dt = \frac{1 + e^{-\pi s}}{s^2 + 1}$$

$$\therefore\ L\left\{\ F(t)\ \right\} = \frac{1 + e^{-\pi s}}{(s^2 + 1)(1 - e^{-2\pi s})}$$

184

$$= \frac{1}{(s^2+1)(1-e^{-\pi s})}$$

Example 5
Find the Laplace transform of the full-wave rectification | sin t | of the sine function.

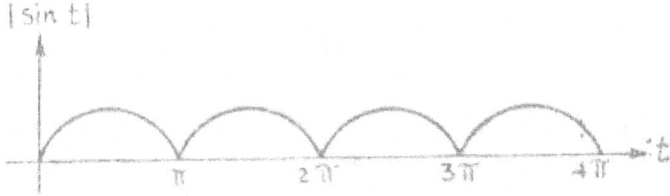

It is evident that in the full wave rectification the ordinates are the sum of the ordinates of F(t) of example 4 + ordinates of F(t) displaced to the right a distance π.

$$\text{Hence } L\left\{ |\sin t| \right\} = \frac{1}{(s^2+1)(1-e^{-\pi s})} + \frac{e^{-\pi s}}{(s^2+1)(1-e^{-\pi s})}$$

$$= \frac{1+e^{-\pi s}}{(s^2+1)(1-e^{-\pi s})}$$

$$= \frac{1}{s^2+1} \cdot \frac{e^{\frac{\pi s}{2}} + e^{-\frac{\pi s}{2}}}{e^{\frac{\pi s}{2}} - e^{-\frac{\pi s}{2}}}$$

$$= \frac{1}{s^2+1} \coth \frac{\pi s}{2}$$

2.13. The product theorem—Convolution
This theorem aims at finding the inverse transform of the product of two transforms. Let $F_1(t)$ and $F_2(t)$ be two sectionally continuous functions in every finite interval $0 \leq t \leq T$ and of the order of e^{at} as $t \longrightarrow \infty$. Take $s > a$ and let

$$L\left\{F_1(t)\right\} = f_1(s), \quad L\left\{F_2(t)\right\} = f_2(s)$$

then
$$L^{-1}\left\{f_1(s)\ f_2(s)\right\} = F_1(t)^* F_2(t) \text{ where}$$

$$F_1(t)^* F_2(t) = \int_0^t F_1(t-\lambda)\ F_2(\lambda)\ d\lambda$$

F1(t)* F2(t) as defined above is called the convolution of the two functions and the last integral is known as a <u>convolution integral</u> or Faltung integral,

Proof :

$$f_1(s) = \int_0^\infty e^{-sx}\ F_1(x)\ dx ,$$

$$f_2(s) = \int_0^\infty e^{-sy}\ F_2(y)\ dy ,$$

$$f_1(s)\ f_2(s) = \int_0^\infty e^{-sx}\ F_1(x)\ dx \int_0^\infty e^{-sy}\ F_2(y)\ dy$$

$$= \int_0^\infty \int_0^\infty e^{-s\,(x+y)}\ F_1(x)\, F_2(y)\ dx\, dy \quad .. \quad (1)$$

FIELD OF INTEGRATION

x-y plane t-λ plane

the integration *being* carried over the positive quadrant
$x > 0, y > 0$.

186

Now let us make the transformation of coordinates

$$
\left.\begin{array}{l} x = t - \lambda \\ y = \lambda \end{array}\right\} \quad \text{i.e} \quad \left.\begin{array}{l} t = x + y \\ \lambda = y \end{array}\right\}
$$

Since $x \geq o$, then $t - \lambda \geq o$ i.e $t \geq \lambda$, Hence the field of integration in the $x - y$ plane transforms in the $t - \lambda$ plane into the field between the positive $t -$ axis and the straight line $\lambda = t$.

Now the element of area dx dy transforms into

$$
\left| \begin{array}{cc} \dfrac{\partial x}{\partial t} & \dfrac{\partial y}{\partial t} \\[2mm] \dfrac{\partial x}{\partial \lambda} & \dfrac{\partial y}{\partial \lambda} \end{array} \right| \; dt\, d\lambda = \left| \begin{array}{cc} 1 & 0 \\ -1 & 1 \end{array} \right| \; dt\, d\lambda = dt\, d\lambda
$$

Hence the double integral (1) transforms into

$$
f_1(s)\, f_2(s) = \int_0^{\infty} \int_0^{t} e^{-st}\, F_1(t-\lambda)\, F_2(\lambda)\, dt\, d\lambda
$$

$$
= \int_0^{\infty} e^{-st} \left[\int_0^{t} F_1(t-\lambda)\, F_2(\lambda)\, d\lambda \right] dt
$$

$$
= L \left\{ \int_0^{t} F_1(t-\lambda)\, F_2(\lambda)\, d\lambda \right\} = L \left\{ F_1 {}^* F_2 \right\}
$$

$$
= L \left\{ F_2 {}^* F_1 \right\} \text{ by symmetry}
$$

Hence we have the following result:

If $\quad L^{-1} \left\{ f_1(s) \right\} = F_1(t)$

and $\quad L^{-1} \left\{ f_2(s) \right\} = F_2(t)$

then $\quad L^{-1} \left\{ f_1(s)\, f_2(s) \right\} = F_1(t)^* F_2(t) = F_2(t)^* F_1(t)$

$$= \int_0^t F_1(t-\lambda)\, F_2(\lambda)\, d\lambda = \int_0^t F_2(t-\lambda)\, F_1(\lambda) d\lambda$$

This formula is known in operational calculus as the <u>Borel formula</u>.

Example 1

$$L^{-1}\left\{\frac{1}{s^2-a^2}\right\} = L^{-1}\left\{\frac{1}{s-a}\cdot\frac{1}{s+a}\right\} = e^{at}\cdot e^{-at}$$

$$= \int_0^t e^{a(t-\lambda)}\, e^{-a\lambda}\, d\lambda = e^{at}\int_0^t e^{-2a\lambda}\, d\lambda = \frac{e^{at}}{-2a}\left[e^{-2a\lambda}\right]_0^t$$

$$= -\frac{e^{at}}{2a}\left[e^{-2at}-1\right] = -\frac{1}{2a}\left[e^{at}-e^{-at}\right] = \frac{1}{a}\sinh at$$

Example 2

$$L^{-1}\left\{\frac{s}{(s^2+a^2)^2}\right\} = L^{-1}\left\{\frac{s}{s^2+a^2}\cdot\frac{1}{s^2+a^2}\right\} = \cos at \cdot \frac{1}{a}\sin at$$

$$= \frac{1}{a}\int_0^t \cos a(t-\lambda)\sin a\lambda\, d\lambda = \frac{1}{2a}\int_0^t [\sin at - \sin a(t-2\lambda)]d\lambda$$

$$= \frac{1}{2a}\left[\lambda \sin at + \frac{\cos a(t-2\lambda)}{-2a}\right]_0^t = \frac{1}{2a}\, t \sin at$$

2.14. <u>Application</u> of the product theorem to the solution of differential and integral equations

Example 1

Solve the differential equation

$Y''(t) + k^2 Y(t) = F(t)$, where k is a constant.

Solution:

$$s^2 y(s) - sY(o) - Y'(o) + k^2 y(s) = f(s)$$

Put $Y(o) = A$ and $Y'(o) = B$

188

$$(s^2+k^2)\; y(s) = As + B + f(s)$$

$$\therefore\; y(s) = \frac{As}{s^2+k^2} + \frac{B}{s^2+k^2} + \frac{1}{s^2+k^2}\cdot f(s)$$

$$\therefore\; Y(t) = A\cos kt + \frac{B}{k}\sin kt + \frac{1}{k}\sin kt^{*}\, F(t)$$

$$= A\cos kt + \frac{B}{k}\sin kt + \frac{1}{k}\int_{0}^{t}\sin k(t-\lambda)\, F(\lambda)\, d\lambda$$

This method is particularly useful in the case when the transform of F(t) is difficult to obtain as in the case of an impressed voltage given for example by an oscillogram. The convolution integral can, however, be calculated numerically.

Example 2

Solve the integral equation

$$Y(t) = 4\sin t + \int_{0}^{t}\sin(t-\lambda)\, Y(\lambda)\, d\lambda \quad \cdots \quad (1)$$

By an integral equation we mean an equation in which the unknown function lies under the integral sign. Here the integral is of the convolution type. Integrals of the convolution type appear in many practical problems. Equation (1) can he written in the form

$Y(t) = 4 \sin t + \sin t^* \, Y(t)$

Transforming we get

$$y(s) = \frac{4}{s^2+1} + \frac{1}{s^2+1}\, y(s)$$

$$\therefore\; y(s)\left[1 - \frac{1}{s^2+1}\right] = \frac{4}{s^2+1}$$

$$\text{i.e }\; y(s) = \frac{4}{s^2} \qquad \therefore\; Y(t) = 4t$$

Example 3
Solve the integral equation

189

$$Y(t) = \tfrac{1}{2} t^2 - \int_0^t (t-\lambda) \, Y(\lambda) \, d\lambda$$

$$\therefore \quad Y(t) = \tfrac{1}{2} t^2 - t^* \, Y(t)$$

$$y(s) = \frac{1}{s^3} - \frac{1}{s^2} \, y(s)$$

$$y(s) \left[1 + \frac{1}{s^2} \right] = \frac{1}{s^3}$$

$$\therefore \quad y(s) \cdot \frac{s^2+1}{s^2} = \frac{1}{s^3}$$

$$\text{i.e} \quad y(s) = \frac{1}{s(s^2+1)} = \frac{1}{s} - \frac{s}{s^2+1}$$

$$\therefore \quad Y(t) = 1 - \cos t$$

Example 4

Solve the integro- differential equation

$$Y'(t) = t + \int_0^t Y(t-\lambda) \cos \lambda \, d\lambda$$

given that $Y(0) = 6$.

Solution:

$$Y'(t) = t + Y(t)^* \cos t$$

$$sy(s) - 6 = \frac{1}{s^2} + y(s) \cdot \frac{s}{s^2+1}$$

$$y(s) \left[s - \frac{s}{s^2+1} \right] = 6 + \frac{1}{s^2}$$

$$y(s) \frac{s^3}{s^2+1} = 6 + \frac{1}{s^2}$$

$$y(s) = \frac{6(s^2+1)}{s^3} + \frac{s^2+1}{s^5}$$

$$= \frac{6}{s} + \frac{7}{s^2} + \frac{1}{s^5}$$

$$\therefore \ Y(t) = 6 + \frac{7}{2} t^2 + \frac{1}{24} t^4$$

2.15. Power series method for the determination of transforms and inverse transforms

In many problems expansion into infinite series is helpful in finding Laplace transforms and inverse transforms. The following are additional examples on the same principle. A list of well known expansions is given here for reference.

(1) $\dfrac{1}{1-x} = \displaystyle\sum_{n=0}^{\infty} x^n$ \qquad $|x| < 1$

(2) $\dfrac{1}{1+x} = \displaystyle\sum_{n=0}^{\infty} (-1)^n x^n$ \qquad $|x| < 1$

(3) $e^x = \displaystyle\sum_{n=0}^{\infty} \dfrac{x^n}{n!}$ \qquad all x

(4) $\cos x = \displaystyle\sum_{n=0}^{\infty} \dfrac{(-1)^n x^{2n}}{(2n)!}$ \qquad all x

(5) $\sin x = \displaystyle\sum_{n=0}^{\infty} \dfrac{(-1)^n x^{2n+1}}{(2n+1)!}$ \qquad all x

(6) $\cosh x = \displaystyle\sum_{n=0}^{\infty} \dfrac{x^{2n}}{(2n)!}$ \qquad all x

(7) $\sinh x = \displaystyle\sum_{n=0}^{\infty} \dfrac{x^{2n+1}}{(2n+1)!}$ \qquad all x

(8) $\tan^{-1} x = \displaystyle\sum_{n=0}^{\infty} \dfrac{(-1)^n x^{2n+1}}{2n+1}$ \qquad $|x| \leqslant 1$

(9) $\log(1+x) = \displaystyle\sum_{n=0}^{\infty} \dfrac{(-1)^n x^{n+1}}{n+1}$ \qquad $-1 < x \leqslant 1$

(10) $\log\dfrac{1+x}{1-x} = 2\displaystyle\sum_{n=0}^{\infty} \dfrac{x^{2n+1}}{2n+1}$ \qquad $|x| < 1$

Example 1

Evaluate $F(t) = L^{-1}\left\{\dfrac{1}{s^3 \cosh 2s}\right\}$ and compute $F(12)$

Solution:

$$\frac{1}{s^3 \cosh 2s} = \frac{2}{s^3\left(e^{2s} + e^{-2s}\right)} = \frac{2e^{-2s}}{s^3\left(1 + e^{-4s}\right)}$$

$$= \sum_{n=0}^{\infty} \frac{2}{s^3} e^{-2s} (-1)^n e^{-4ns}$$

$$= \sum_{n=0}^{\infty} \frac{2}{s^3} (-1)^n e^{(-4n-2))s}$$

$$L^{-1}\left\{\frac{1}{s^3 \cos h\, 2s}\right\} = \sum_{n=0}^{\infty} (-1)^n (t-4n-2)^2\, U(t-4n-2)$$

$$\therefore\ F(t) = (t-2)^2\, U(t-2) - (t-6)^2\, U(t-6) + (t-10)^2\, U(t-10)$$
$$- (t-14)^2\, U\,(t-14) + \dots$$

Notice that the series in the right hand side terminates once the argument of the function U is negative. Hence for $t = 12$ we have

$$F(t) = (t-2)^2\, U(t-2) - (t-6)^2\, U(t-6) + (t-10)^2\, U(t-10)$$

$$\therefore\quad F(12) = 100 - 36 + 4 = 68$$

Example 2.

Evaluate $L^{-1}\left\{\log \dfrac{s+1}{s-1}\right\}$

Solution:

$$\log \frac{s+1}{s-1} = \log \frac{1 + \dfrac{1}{s}}{1 - \dfrac{1}{s}} = 2 \sum_{n=0}^{\infty} \frac{1}{(2n+1)\, s^{2n+1}}$$

(from series 10)

Now $L^{-1}\left\{\dfrac{1}{s^{2n+1}}\right\} = \dfrac{t^{2n}}{(2n)!}$, Hence

$$L^{-1}\left\{\log \dfrac{s+1}{s-1}\right\} = 2\sum_{n=0}^{\infty}\dfrac{t^{2n}}{(2n+1)!} = \dfrac{2}{t}\sum_{n=0}^{\infty}\dfrac{t^{2n+1}}{(2n+1)!}$$

$$= \dfrac{2}{t}\sinh t \text{ (from series 7)}$$

Example 3

Find $L\left\{\displaystyle\int_{0}^{t}\dfrac{\sin u}{u}\,du\right\}$

Solution:

$$\int_{0}^{t}\dfrac{\sin u}{u}\,du = \int_{0}^{t}\dfrac{1}{u}\left[u - \dfrac{u^3}{3!} + \dfrac{u^5}{5!} - \cdots\right]du$$

$$= t - \dfrac{t^3}{3.3!} + \dfrac{t^5}{5.5!} - \cdots$$

$$\therefore\ L\left\{\int_{0}^{t}\dfrac{\sin u}{u}\,du\right\} = \dfrac{1}{s^2} - \dfrac{1}{3.3!}\dfrac{3!}{s^4} + \dfrac{1}{5.5!}\cdot\dfrac{5!}{s^6} - \cdots$$

$$= \dfrac{1}{s^2} - \dfrac{1}{3s^4} + \dfrac{1}{5s^6} - \cdots$$

$$= \dfrac{1}{s}\left\{\dfrac{(1/s)}{1} - \dfrac{(1/s)^3}{3} + \dfrac{(1/s)^5}{5}\right\} - \cdots$$

$$= \dfrac{1}{s}\tan^{-1}\dfrac{1}{s}$$

Example 4
If $s > o$ and $n > 1$ prove that

$$L\left\{\dfrac{t^{n-1}}{1-e^{-t}}\right\} = \Gamma(n)\left\{\dfrac{1}{s^n} + \dfrac{1}{(s+1)^n} + \dfrac{1}{(s+2)^n} + \cdots\right\}$$

Solution:

$$\frac{t^{n-1}}{1-e^{-t}} = t^{n-1}(1+e^{-t}+e^{-2t}+\cdots)$$

$$= t^{n-1} + t^{n-1}e^{-t} + t^{n-1}e^{-2t} + \cdots$$

$$\therefore \quad L\left\{\frac{t^{n-1}}{1-e^{-t}}\right\} = \frac{\Gamma(n)}{s^n} + \frac{\Gamma(n)}{(s+1)^n} + \frac{\Gamma(n)}{(s+2)^n} + \cdots$$

$$= \Gamma(n)\left\{\frac{1}{s^n} + \frac{1}{(s+1)^n} + \frac{1}{(s+2)^n} + \cdots\right\}$$

Example 5

Find $L^{-1}\left\{\dfrac{e^{-1/s}}{s}\right\}$

Solution:

$$\frac{1}{s}e^{-1/s} = \frac{1}{s}\left\{1 - \frac{1}{s} + \frac{1}{2!s^2} - \frac{1}{3!s^3} + \cdots\right\}$$

$$= \frac{1}{s} - \frac{1}{s^2} + \frac{1}{2!s^3} - \frac{1}{3\,s^4} + \cdots$$

$$L^{-1}\left\{\frac{1}{s}e^{-1/s}\right\} = 1 - t + \frac{t^2}{(2!)^2} - \frac{t^3}{(3!)^2} + \cdots$$

$$= 1 - t + \frac{t^2}{1^2 . 2^2} - \frac{t^3}{1^2 . 2^2 . 3^2} + \cdots$$

$$= 1 - \frac{(2t^{\frac{1}{2}})^2}{2^2} + \frac{(2t^{\frac{1}{2}})^4}{2^2 . 4^2} - \frac{(2t^{\frac{1}{2}})^6}{2^2 . 4^2 . 6^2} + \cdots$$

$$= J_0(2\sqrt{t})$$

2.16. The error function or probability integral

This is a tabulated function denoted by erf(x) and is defined as follows

$$\text{erf}(x) = \frac{2}{\sqrt{\pi}} \int_0^x e^{-u^2} \, du$$

$$= \frac{2}{\sqrt{\pi}} \times \text{shaded area}$$

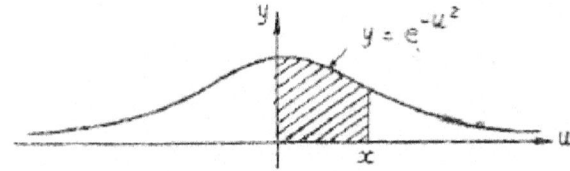

Expanding e^{-u^2} in ascending powers of u and integrating term by term we get

$$\text{erf}(x) = \frac{2}{\sqrt{\pi}} \int_0^x \left(1 - u^2 + \frac{u^4}{2!} - \frac{u^6}{3!} + \cdots \right) du$$

$$= \frac{2}{\sqrt{\pi}} \left[x - \frac{x^3}{3} + \frac{x^5}{2!5} - \frac{x^7}{3!7} + \cdots \right]$$

Now since $\int_0^\infty e^{-u^2} \, du = \frac{\sqrt{\pi}}{2}$, it follows that $\text{erf}(x) \longrightarrow 1$

as $x \longrightarrow \infty$.

The complementary error function erfc (x) is defined as follows

$$\text{erfc}(x) = 1 - \text{erf}(x) = \frac{2}{\sqrt{\pi}} \int_0^\infty e^{-u^2} \, du - \frac{2}{\sqrt{\pi}} \int_0^x e^{-u^2} \, du$$

$$= \frac{2}{\sqrt{\pi}} \int_x^\infty e^{-u^2} \, du$$

Example 1.

Find $L^{-1} \left\{ \dfrac{1}{(s-1)\sqrt{s}} \right\}$

Solution:

We have shown that

$$L \left\{ t^{-\frac{1}{2}} \right\} = \sqrt{\dfrac{\pi}{s}}$$

Hence $L^{-1} \left\{ \dfrac{1}{\sqrt{s}} \right\} = \dfrac{1}{\sqrt{\pi t}}$

$$\therefore \quad L^{-1} \left\{ \dfrac{1}{(s-1)\sqrt{s}} \right\} = e^t \cdot \dfrac{1}{\sqrt{\pi t}} = \dfrac{1}{\sqrt{\pi}} \int_0^t e^{t-\lambda} \dfrac{1}{\sqrt{\lambda}} d\lambda$$

$$= \dfrac{e^t}{\sqrt{\pi}} \int_0^t e^{-\lambda} \dfrac{1}{\sqrt{\lambda}} d\lambda$$

Put $\lambda = u^2$, then $\displaystyle\int_0^t e^{-\lambda} \dfrac{1}{\sqrt{\lambda}} d\lambda = \int_0^{\sqrt{t}} \dfrac{e^{-u^2}}{u} \cdot 2u\, du = 2 \int_0^{\sqrt{t}} e^{-u^2} du$

$$\therefore \quad L^{-1} \left\{ \dfrac{1}{(s-1)\sqrt{s}} \right\} = e^t \times \dfrac{2}{\sqrt{\pi}} \int_0^{\sqrt{t}} e^{-u^2} du = e^t \, \text{erf}(\sqrt{t})$$

If we change s into s +1 we get

$$L^{-1} \left\{ \dfrac{1}{s\sqrt{s+1}} \right\} = e^{-t} \cdot e^t \, \text{erf}(\sqrt{t}) = \text{erf}(\sqrt{t})$$

Example 2

Find $L \left\{ e^{-2t} \, \text{erf}(\sqrt{t}) \right\}$

Solution:

Since $L\left\{erf(\sqrt{t})\right\} = \dfrac{1}{s\sqrt{s+1}}$, then

$$L\left\{e^{2t}\,erf(\sqrt{t})\right\} = \dfrac{1}{(s-2)\sqrt{s-1}}$$

Example 3

Find $L\left\{t\,erf(2\sqrt{t})\right\}$

Solution:

Since $L\left\{erf(\sqrt{t})\right\} = \dfrac{1}{s\sqrt{s+1}}$, then

$$L\left\{erf(2\sqrt{t})\right\} = L\left\{erf(\sqrt{4t})\right\}$$

$$= \dfrac{1}{4} \cdot \dfrac{1}{\dfrac{s}{4}\sqrt{\dfrac{s}{4}+1}} = \dfrac{2}{s\sqrt{s+4}}$$

$$\therefore L\left\{t\,erf(2\sqrt{t})\right\} = -\dfrac{d}{ds}\dfrac{2}{s\sqrt{s+4}} = \dfrac{3s+8}{s^2(s+4)^{3/2}}$$

Example 4

Find $L\left\{erfc(\sqrt{t})\right\}$

Solution:

Since $erfc(\sqrt{t}) = 1 - erf(\sqrt{t})$, then

$$L\left\{erfc(\sqrt{t})\right\} = \dfrac{1}{s} - \dfrac{1}{s\sqrt{s+1}} = \dfrac{\sqrt{s+1}-1}{s\sqrt{s+1}}$$

$$= \dfrac{1}{\sqrt{s+1}[\sqrt{s+1}+1]}$$

Example 5

Find $L^{-1}\left\{\dfrac{1}{1+\sqrt{1+s}}\right\}$

Solution:

$$\frac{1}{1+\sqrt{1+s}} = \frac{1-\sqrt{1+s}}{1-(1+s)} = \frac{1-\sqrt{1+s}}{-s}$$

$$= -\frac{1}{s} + \frac{\sqrt{1+s}}{s} = -\frac{1}{s} + \frac{1+s}{s\sqrt{1+s}}$$

$$= -\frac{1}{s} + \frac{1}{s\sqrt{s+1}} + \frac{1}{\sqrt{s+1}}$$

$$\therefore \quad L^{-1}\left\{\frac{1}{1+\sqrt{1+s}}\right\} = -1 + \mathrm{erf}(\sqrt{t}) + \frac{e^{-t}}{\sqrt{\pi t}}$$

$$= \frac{e^{-t}}{\sqrt{\pi t}} - \mathrm{erfc}(\sqrt{t})$$

Example 6

Find $L\left\{\displaystyle\int_0^t \mathrm{erf}(\sqrt{u})\,du\right\}$

Solution:

Since the integration of the object function between the limits 0 and t corresponds to the division of the transform by s, then

$$L\left\{\int_0^t \mathrm{erf}(\sqrt{u})\,du\right\} = \frac{1}{s} \cdot \frac{1}{s\sqrt{s+1}} = \frac{1}{s^2\sqrt{s+1}}$$

2.17. The sine-integral function Si(t)

This is a tabulated function defined by

$$Si(t) = \int_0^t \frac{\sin u}{u}\,du$$

It is represented by the shaded area under the curve $y = \dfrac{\sin u}{u}$ between o and t.

198

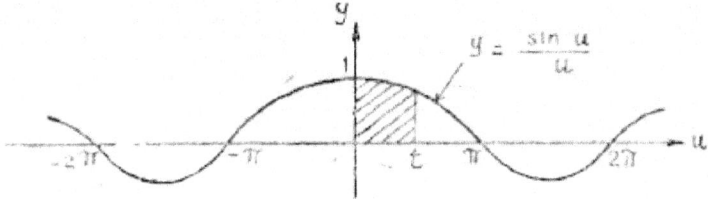

The function (sin u)/u is an even function which → 0 as u → ∞.
It intersects the u — axis where

$$u = \pm \pi, \pm 2\pi, \pm 3\pi, \ldots$$

Now $L\left\{\sin t\right\} = \dfrac{1}{s^2 + 1}$

∴ $L\left\{\dfrac{\sin t}{t}\right\} = \int_s^\infty \dfrac{dx}{x^2 + 1} = \left[\tan^{-1} x\right]_s^\infty$

$$= \dfrac{\pi}{2} - \tan^{-1} s = \cot^{-1} s$$

∴ $L\left\{\int_0^t \dfrac{\sin u}{u}\, du\right\} = \dfrac{1}{s} \cot^{-1} s$

i.e $L\left\{Si(t)\right\} = \dfrac{1}{s} \cot^{-1} s$, $s > 0$,

The following is an alternative method for finding L {Si (t)}

Put $F(t) = \int_0^t \dfrac{\sin u}{u}\, du$

then $F'(t) = \dfrac{\sin t}{t}$ or $t\, F'(t) = \sin t$

∴ $L\left\{t\, F'(t)\right\} = L\left\{\sin t\right\}$

∴ $-\dfrac{d}{ds}\left[s\, f(s) - F(o)\right] = \dfrac{1}{s^2 + 1}$

∴ $\dfrac{d}{ds}\left[s\, f(s)\right] = -\dfrac{1}{s^2 + 1}$

Integrating

$$s\,f(s) = -\tan^{-1}s + C$$

According to the initial value theorem

$$\lim_{t \to 0} F(t) = \lim_{s \to \infty} s\,f(s)$$

and since $\lim_{t \to 0} F(t) = F(o) = o$, then $\lim_{s \to \infty} s\,f(s) = o$

$$\therefore \quad -\frac{\pi}{2} + C = o \quad \text{i.e} \quad C = \frac{\pi}{2} \quad \text{and hence}$$

$$s\,f(s) = \frac{\pi}{2} - \tan^{-1}s = \cot^{-1}s$$

$$\therefore \quad f(s) = \frac{1}{s}\cot^{-1}s \quad \text{i.e} \quad L\left\{Si(t)\right\} = \frac{1}{s}\cot^{-1}s$$

2.18. The Cosine -integral function Ci(t)

This is a tabulated function defined by

$$Ci(t) = -\int_{t}^{\infty} \frac{\cos u}{u}\,du , \qquad t > o .$$

change the variable u into a new variable x by the relation u = xt where t is a constant, then

$$Ci(t) = -\int_{1}^{\infty} \frac{\cos xt}{xt}\,t\,dx = -\int_{1}^{\infty} \frac{\cos xt}{x}\,dx$$

$$\therefore \quad L\left\{Ci(t)\right\} = -\int_{0}^{\infty} e^{-st}\left\{\int_{1}^{\infty} \frac{\cos xt}{x}\,dx\right\}\,dt$$

$$= -\int_{1}^{\infty} \frac{1}{x}\left\{\int_{0}^{\infty} e^{-st}\cos xt\,dt\right\}\,dx$$

(by interchanging the order of integration)

$$= -\int_{1}^{\infty} \frac{1}{x}\left[\frac{s}{s^2 + x^2}\right]\,dx = -\frac{s}{s^2}\int_{1}^{\infty}\left[\frac{1}{x} - \frac{x}{s^2 + x^2}\right]\,dx$$

$$= -\frac{1}{s}\left[\log x - \tfrac{1}{2}\log(s^2 + x^2)\right]_{1}^{\infty}$$

$$= -\frac{1}{s}\left[\log\frac{x}{\sqrt{s^2+x^2}}\right]_1^\infty$$

$$= \frac{1}{s}\log\frac{1}{\sqrt{s^2+1}} = -\frac{1}{2s}\log(s^2+1)$$

$$\therefore\ L\left\{Ci(t)\right\} = -\frac{1}{2s}\log(s^2+1),\qquad s>0$$

The transform can also be obtained in the following *way:*

$$\text{Put}\ F(t) = -\int_t^\infty\frac{\cos u}{u}\,du\,,\qquad\text{then}$$

$$F'(t) = \frac{\cos t}{t}\qquad\qquad\text{i.e}\qquad t\,F'(t) = \cos t$$

$$-\frac{d}{ds}\left[s\,f(s) - F(o)\right] = \frac{s}{s^2+1}$$

$$\therefore\ \frac{d}{ds}\left[s\,f(s)\right] = -\frac{s}{s^2+1}$$

Integrating

$$s\,f(s) = -\tfrac{1}{2}\log(s^2+1) + C$$

And according to the final value theorem

$$\lim_{t\to\infty}F(t) = \lim_{s\to0}s\,f(s)$$

$$\therefore\ 0 = \lim_{s\to0}s\,f(s) = C$$

$$\therefore\ s\,f(s) = -\tfrac{1}{2}\log(s^2+1)$$

$$\text{i.e}\quad f(s) = -\frac{1}{2s}\log(s^2+1)$$

$$\therefore\ L\left\{Ci(t)\right\} = -\frac{1}{2s}\log(s^2+1)$$

2.19. The exponential integral function

This is defined by

201

$$Ei(t) = \int_{-\infty}^{t} \frac{e^u}{u} \, du \qquad (t < o)$$

This can be written

$$- Ei(-t) = \int_{t}^{\infty} \frac{e^{-y}}{y} \, dy = \int_{1}^{\infty} \frac{e^{-tx}}{x} \, dx \qquad (t > o)$$

the last integral being obtained by the substitution $y = tx$ where t is a constant.

$$L \left\{ \int_{1}^{\infty} \frac{e^{-tx}}{x} \, dx \right\} = \int_{0}^{\infty} e^{-st} \left\{ \int_{1}^{\infty} \frac{e^{-tx}}{x} \, dx \right\} dt$$

$$= \int_{1}^{\infty} \frac{1}{x} \left\{ \int_{0}^{\infty} e^{-st} \, e^{-tx} \, dt \right\} dx = \int_{1}^{\infty} \frac{1}{x} \cdot \frac{1}{s+x} \, dx$$

$$= \frac{1}{s} \int_{1}^{\infty} \left[\frac{1}{x} - \frac{1}{s+x} \right] dx = \frac{1}{s} \left[\log \frac{x}{s+x} \right]_{1}^{\infty}$$

$$= - \frac{1}{s} \log \frac{1}{s+1} = \frac{1}{s} \log (s+1)$$

$$\therefore \quad L \left\{ - Ei(-t) \right\} = \frac{1}{s} \log (s+1), \qquad s > o$$

2.20. Evaluation of definite integrals using the Laplace transformation

Example 1

Evaluate $\quad F(t) = \int_{0}^{\infty} \frac{\sin tx}{x(x^2 + 1)} \, dx, \qquad t > a$

Solution:

$$L \left\{ F(t) \right\} = \int_{0}^{\infty} e^{-st} \left\{ \int_{0}^{\infty} \frac{\sin tx}{x(x^2 + 1)} \, dx \right\} dt$$

$$= \int_{0}^{\infty} \frac{1}{x(x^2+1)} \left\{ \int_{0}^{\infty} e^{-st} \sin tx \, dt \right\} dx$$

202

(by interchanging the order of integration)

$$= \int_0^\infty \frac{1}{x(x^2+1)} \cdot \frac{x}{s^2+x^2} \, dx = \int_0^\infty \frac{1}{(x^2+1)(x^2+s^2)} \, dx$$

$$= \frac{1}{s^2-1} \int_0^\infty \left[\frac{1}{x^2+1} - \frac{1}{x^2+s^2} \right] \, dx$$

$$= \frac{1}{s^2-1} \left[\tan^{-1} x - \frac{1}{s} \tan^{-1} \frac{x}{s} \right]_0^\infty$$

$$= \frac{1}{s^2-1} \left(\frac{\pi}{2} - \frac{1}{s} \frac{\pi}{2} \right) = \frac{\pi}{2} \cdot \frac{1}{s^2-1} \left(1 - \frac{1}{s} \right)$$

$$= \frac{\pi}{2} \cdot \frac{1}{s(s+1)} = \frac{\pi}{2} \left[\frac{1}{s} - \frac{1}{s+1} \right]$$

$$\therefore \quad F(t) = \frac{\pi}{2}[1-e^{-t}].$$

Example 2

Evaluate $\displaystyle\int_0^\infty e^{-x^2} \, dx$

Solution:

Consider $\displaystyle F(t) = \int_0^\infty e^{-tx^2} \, dx \ , \ t > 0$

$$L\left\{ F(t) \right\} = \int_0^\infty e^{-st} \left\{ \int_0^\infty e^{-tx^2} dx \right\} \, dt$$

$$= \int_0^\infty \left\{ \int_0^\infty e^{-st} \cdot e^{-tx^2} \, dt \right\} dx$$

$$= \int_0^\infty \frac{1}{s+x^2} \, dx = \left[\frac{1}{\sqrt{s}} \tan^{-1} \frac{x}{\sqrt{s}} \right]_0^\infty = \frac{1}{\sqrt{s}} \cdot \frac{\pi}{2}$$

$$\therefore \quad F(t) = \frac{\pi}{2} L^{-1}\left\{ \frac{1}{\sqrt{s}} \right\} = \frac{\pi}{2} \cdot \frac{1}{\sqrt{\pi t}} \qquad (1.2)$$

$$= \frac{\sqrt{\pi}}{2} \cdot \frac{1}{\sqrt{t}}$$

$$\therefore \int_0^\infty e^{-tx^2} dx = \frac{\sqrt{\pi}}{2} \cdot \frac{1}{\sqrt{t}}$$

Putting t = 1 *we* get

$$\int_0^\infty e^{-x^2} dx = \frac{\sqrt{\pi}}{2}$$

Example 3

$$\text{Evaluate} \int_0^t J_0(t-\lambda) J_0(\lambda) \, d\lambda$$

Solution:

The integral is the convolution $J_0(t) \cdot J_0(t)$ and its transform is

$$\frac{1}{\sqrt{s^2+1}} \cdot \frac{1}{\sqrt{s^2+1}} = \frac{1}{s^2+1}$$

Taking the inverse transform we get

$$\int_0^t J_0(t-\lambda) J_0(\lambda) \, d\lambda = \sin t$$

Example 4

$$\text{Evaluate } F(t) = \int_0^\infty \frac{\cos tx}{\sqrt{x}} \, dx \, , \, t > 0$$

Solution:

$$L\left\{F(t)\right\} = \int_0^\infty e^{-st} \left\{ \int_0^\infty \frac{\cos tx}{\sqrt{x}} \, dx \right\} dt$$

$$= \int_0^\infty \frac{1}{\sqrt{x}} \left\{ \int_0^\infty e^{-st} \cos tx \, dt \right\} dx$$

204

$$= \int_0^\infty \frac{1}{\sqrt{x}} \cdot \frac{s}{s^2 + x^2} \, dx$$

Put $x = s \tan \Theta$ $\therefore$ $dx = s \sec^2\Theta \, d\Theta$

$$\therefore \quad f(s) = \int_0^{\frac{\pi}{2}} \frac{1}{\sqrt{s \tan\Theta}} \cdot \frac{s}{s^2 \sec^2\Theta} \cdot s \sec^2\Theta \, d\Theta$$

$$= \frac{1}{2\sqrt{s}} \int_0^{\frac{\pi}{2}} 2 \cos^{\frac{1}{2}}\Theta \, \sin^{-\frac{1}{2}}\Theta \, d\Theta$$

$$= \frac{1}{2\sqrt{s}} \beta\left(\frac{1}{4}, \frac{3}{4}\right) = \frac{1}{2\sqrt{s}} \frac{\Gamma(1/4) \, \Gamma(3/4)}{\Gamma(1)}$$

$$= \frac{1}{2\sqrt{s}} \frac{\pi}{\sin \frac{\pi}{4}}, \text{ since } \Gamma(m)\,\Gamma(1-m) = \frac{\pi}{\sin m\pi} \quad (0 < m < 1)$$

$$= \frac{\pi}{\sqrt{2}} \cdot \frac{1}{\sqrt{s}}$$

$$\therefore \quad F(t) = \frac{\pi}{\sqrt{2}} \cdot \frac{1}{\sqrt{\pi t}} = \sqrt{\frac{\pi}{2t}}$$

$$\therefore \quad \int_0^\infty \frac{\cos tx}{\sqrt{x}} \, dx = \sqrt{\frac{\pi}{2t}}$$

Example 5

Evaluate $\displaystyle\int_0^\infty \sin x^2 \, dx$

Solution:

Let $\displaystyle F(t) = \int_0^\infty \sin tx^2 \, dx$, $\quad t > 0$

$$\therefore \quad L\left\{F(t)\right\} = \int_0^\infty e^{-st} \left\{ \int_0^\infty \sin tx^2 \, dx \right\} dt$$

205

$$= \int_0^\infty \left\{ \int_0^\infty e^{-st} \sin tx^2 \, dt \right\} dx$$

$$= \int_0^\infty \frac{x^2}{s^2 + x^4} \, dx$$

Put $x^2 = s \tan \theta$ $\therefore$ $2xdx = s \sec^2\theta \, d\theta$

when $x = o$, $\theta = o$ and when $x \longrightarrow \infty$, $\theta \longrightarrow \dfrac{\pi}{2}$

$$\text{Integral} = \tfrac{1}{2} \int_0^{\frac{\pi}{2}} \frac{\sqrt{s \tan \theta} \; s \sec^2\theta \, d\theta}{s^2 \sec^2\theta}$$

$$= \frac{1}{2\sqrt{s}} \int_0^{\frac{\pi}{2}} \sqrt{\tan \theta} \, d\theta = \frac{1}{2\sqrt{s}} \int_0^{\frac{\pi}{2}} \sin^{\frac{1}{2}}\theta \, \cos^{-\frac{1}{2}} \theta \, d\theta$$

$$= \frac{1}{4\sqrt{s}} \int_0^{\frac{\pi}{2}} 2 \sin^{\frac{1}{2}}\theta \, \cos^{-\frac{1}{2}}\theta \, d\theta$$

$$= \frac{1}{4\sqrt{s}} \beta \left(\tfrac{1}{4}, \tfrac{3}{4} \right) = \frac{1}{4\sqrt{s}} \frac{\Gamma(\tfrac{1}{4}) \, \Gamma(\tfrac{3}{4})}{\Gamma(1)}$$

$$= \frac{1}{4\sqrt{s}} \frac{\pi}{\sin \dfrac{\pi}{4}} = \frac{\pi}{2\sqrt{2}\sqrt{s}}$$

inverting we get

$$F(t) = \int_0^\infty \sin tx^2 \, dx = \frac{\pi}{2\sqrt{2}} \cdot \frac{1}{\sqrt{\pi t}} = \frac{\sqrt{\pi}}{2\sqrt{2}\sqrt{t}}$$

Put $t = 1$

$$\int_0^\infty \sin x^2 \, dx = \tfrac{1}{2} \sqrt{\frac{\pi}{2}}$$

2.21. The Heaviside's expansion formulae

206

Let $f(s) = p(s)/q(s)$ where $p(s)$ and $q(s)$ are polynomials in s, $p(s)$ being of lower degree than $q(s)$.

Case I : Consider the case in which all factors of $q(s)$ are linear and distinct.

Let $q(s) = (s-a_1)(s-a_2) \ldots (s-a_n)$, the a's being all distinct.

Then $f(s) = \dfrac{p(s)}{q(s)} = \dfrac{A_1}{s-a_1} + \dfrac{A_2}{s-a_2} + \ldots + \dfrac{A_r}{s-a_r}$

$$+ \ldots + \dfrac{A_n}{s-a_n}$$

Clearing fractions we get

$$p(s) = A_1 (s-a_2)(s-a_3) \ldots (s-a_n)$$
$$+ A_2 (s-a_1)(s-a_3) \ldots (s-a_n)$$
$$+ \ldots \ldots \ldots \ldots$$
$$+ A_n (s-a_1)(s-a_2) \ldots (s-a_{n-1})$$

Put $s = a_1$, then

$$p(a_1) = A_1 (a_1-a_2)(a_1-a_3) \ldots (a_1-a_n)$$

$$\therefore A_1 = \dfrac{p(a_1)}{(a_1-a_2)(a_1-a_3) \ldots (a_1-a_n)}$$

Hence A_1 is obtained by substituting a_1 in the numerator and in all factors of the denominator except the factor $s-a_1$ itself. Similarly for $A_2, A_3 \ldots, A_n$.

Now $q'(s) = (s-a_2)(s-a_3) \ldots (s-a_n)$
$$+ (s-a_1)(s-a_3) \ldots (s-a_n)$$
$$+ \ldots \ldots \ldots \ldots$$
$$+ (s-a_1)(s-a_2) \ldots (s-a_{n-1})$$

$$\therefore q'(a_1) = (a_1-a_2)(a_1-a_3) \ldots (a_1-a_n)$$

Hence we can write $A_1 = \dfrac{p(a_1)}{q'(a_1)}$ and in general $A_r = \dfrac{p(a_r)}{q'(a_r)}$

$$\therefore \ f(s) = \frac{p(s)}{q(s)} = \sum_{r=1}^{n} \frac{p(a_r)}{q'(a_r)} \cdot \frac{1}{s-a_r}$$

$$\therefore \ F(t) = L^{-1}\left\{\frac{p(s)}{q(s)}\right\} = \sum_{r=1}^{n} \frac{p(a_r)}{q'(a_r)} e^{a_r t}$$

Case II: $g(s)$ has repeated linear factors.

Suppose that the denominator $q(s)$ has a repeated linear factor $(s-a)^n$ and that

$$\frac{p(s)}{q(s)} = \frac{\phi(s)}{(s-a)^n}$$

Where $\phi(s)$ consists of the numerator $p(s)$ and all factors of $q(s)$ except $(s-a)^n$

$$\phi(s) = \phi\,(a+\overline{s-a}) = \phi(a) + (s-a)\,\phi'(a) + \frac{(s-a)^2}{2!}\phi''(a) + \cdots$$

$$+ \frac{(s-a)^{n-1}}{(n-1)!}\phi^{(n-1)}(a) + (s-a)^n h(s).$$

$$\therefore \ \frac{p(s)}{q(s)} = \frac{\phi(s)}{(s-a)^n} = \frac{\phi(a)}{(s-a)^n} + \frac{\phi'(a)}{(s-a)^{n-1}} + \frac{\phi''(a)}{2!(s-a)^{n-2}} + \cdots$$

$$+ \frac{\phi^{(n-1)}(a)}{(n-1)!}\frac{1}{s-a} + h(s).$$

$$\therefore \ L^{-1}\left\{\frac{p(s)}{q(s)}\right\} = e^{at}\left[\frac{t^{n-1}}{(n-1)!}\phi(a) + \frac{t^{n-2}}{(n-1)!}\phi'(a)\right.$$

$$\left. + \frac{t^{n-3}}{(n-3)!}\frac{\phi''(a)}{!} + \cdots + \frac{\phi^{(n-1)}(a)}{(n-1)!}\right] + H(t)$$

$$= e^{at}\sum_{r=1}^{n} \frac{t^{n-r}}{(n-r)!}\frac{\phi^{(r-1)}(a)}{(r-1)!}$$

Case III: q(s) has quadratic factors.

let $\dfrac{p(s)}{q(s)} = \dfrac{\phi(s)}{(s-a)^2+b^2}$ where $\phi(s)$ consists of p(s)

and all factors of q (s) except the factor $(s-a)^2+b^2$ itself.

Let $\dfrac{\phi(s)}{(s-a)^2+b^2} = \dfrac{As+B}{(s-a)^2+b^2} + h(s)$

$\therefore \quad \phi(s) = As + B + [(s-a)^2+b^2] \, h(s)$

Put s = a + ib

$\therefore \quad \phi(a+ib) = A(a+ib) + B$

Let $\phi(a+ib) = \phi_1 + i\phi_2$

$\therefore \quad \phi_1 + i\phi_2 = Aa + B + iAb$

$\therefore \quad \phi_1 = Aa + B, \; \phi_2 = Ab$

$\therefore \quad A = \dfrac{\phi_2}{b}, \; \phi_1 = \dfrac{a}{b}\phi_2 + B$ i.e $B = \dfrac{b\phi_1 - a\phi_2}{b}$

$$\dfrac{\phi(s)}{(s-a)^2+b^2} = \dfrac{\dfrac{\phi_2}{b}s + \dfrac{b\phi_1 - a\phi_2}{b}}{(s-a)^2+b^2} + h(s)$$

$$= \dfrac{1}{b}\dfrac{\phi_2 s + b\phi_1 - a\phi_2}{(s-a)^2+b^2} + h(s)$$

$$= \dfrac{1}{b}\dfrac{\phi_2(s-a) + b\phi_1}{(s-a)^2+b^2} + h(s)$$

$\therefore \quad L^{-1}\left\{\dfrac{p(s)}{q(s)}\right\} = \dfrac{1}{b}e^{at}[\phi_2 \cos bt + \phi_1 \sin bt] + H(t)$

Case IV: q(s) has repeated quadratic factors.

Consider for example the particular case when

$$\dfrac{p(s)}{q(s)} = \dfrac{\phi(s)}{[(s-a)^2+b^2]^2} = \dfrac{As + B}{[(s-a)^2+b^2]^2} + \dfrac{Cs + D}{(s-a)^2+b^2} + h(s)$$

then $\phi(s) = As+B+(Cs+D)[(s-a)^2+b^2] + h(s)[(s-a)^2+b^2]^2 \ldots (1)$

209

The four unknowns A,B,C,D can be obtained by substituting $s = a + i b$ in (1) and in the equation derived by differentiating (I) w.r.t. s and equating real to real and imaginary to imaginary on both sides in each case.

2.22. The inversion integral

The inversion integral is a powerful as well as a direct mean for finding inverse Laplace transforms. We have so far assumed s to be real. We shall now consider s to be complex. Let F(t) be a real function of the positive, real variable t, sectionally continuous in each finite interval $0 \leq t \leq T$ and of exponential order as $t \rightarrow \infty$

$$f(s) = \int_0^\infty e^{-st} F(t) \, dt \quad \text{where} \quad s = x + iy$$

then the inversion integral formula is given by

$$F(t) = \frac{1}{2\pi i} \lim_{\beta \to \infty} \int_{\gamma - i\beta}^{\gamma + i\beta} e^{st} f(s) \, ds \quad \ldots \quad (1)$$

This formula is also known as <u>Bromwich's integral formula</u>, and it is a modified form of the Fourier integral formula. The integration is carried in the complex plane of s along the line $x = y$ which is taken far enough to the right such that all poles, branch points or essential singularities of $e^{-st} f(s)$ lie to the left of it. In practice, the integral (1) is evaluated by considering first the contour integral

$$\frac{1}{2\pi i} \int_C e^{st} f(s) \, ds$$

where C is the contour consisting of the line AB whose equation is $x = y$ and the circular arc BDA which we denote by Γ of a circle centre O and radius R. R is taken large enough such that all poles of $e^{-st} f(s)$ lie inside C, and in the limit R is made infinite.

Now since $\beta = \sqrt{R^2 - \gamma^2}$, hence when $\beta \rightarrow \infty$, $R \rightarrow \infty$ and according to the inversion formula

$$F(t) = \lim_{R \to \infty} \frac{1}{2\pi i} \int_{\gamma-i\beta}^{\gamma+i\beta} e^{st} f(s)\ ds$$

$$= \lim_{R \to \infty} \left\{ \frac{1}{2\pi i} \int_C e^{st} f(s)\ ds - \frac{1}{2\pi i} \int_T e^{st} f(s)\ ds \right\}$$

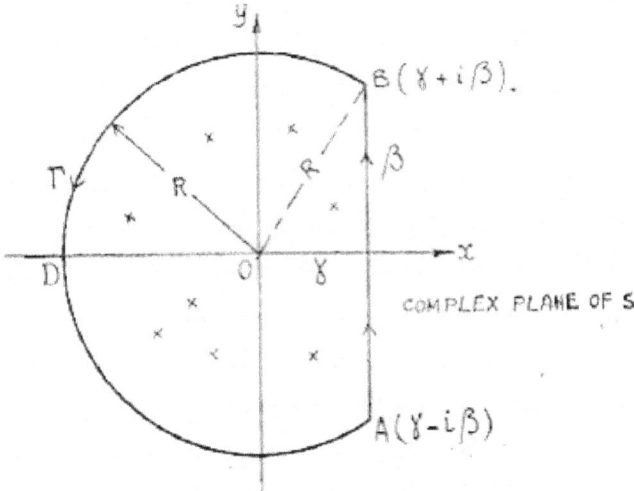

Assuming that all singularities of f(s) are poles and that

$$\int_\Gamma e^{st} f(s)\ ds \to 0 \text{ as } R \to \infty,$$

then, since by Cauchy's theorem of residues

$$\int_C e^{st} f(s)\ ds = 2\pi i$$

(sum of residues of $e^{st} f(s)$ at its poles inside C) it follows that

F(t) = sum of residues of $e^{st} f(s)$ at poles of f(s).

This result was obtained on the assumption that

$$\int_{\Gamma} e^{st} f(s) \, ds \longrightarrow 0 \text{ as } R \longrightarrow \infty.$$

This assumption is satisfied in almost all physical problems. In fact a sufficient condition for the ultimate vanishing of the above integral is the following.

If we can find two positive constants M and k such that on Γ where $s = Re^{i\theta}$

$$|f(s)| < \frac{M}{R^k}$$

then the integral around Γ of $e^{st} f(s) \longrightarrow 0$ as $R \longrightarrow \infty$

$$\text{i.e } \lim_{R \longrightarrow \infty} \int_{\Gamma} e^{st} f(s) \, ds = 0$$

2.23. Formulae for residues

I. Simple poles.

Let

$$\frac{p(s)}{q(s)} = \frac{\phi(s)}{s - s_0}$$

where $\phi(s)$ consists of p(s) and all factors of q(s) except the factor $s - s_0$ itself. Here, we have a simple pole at $s - s_0$.

$$\text{Now } \phi(s) = \phi(s_0 + s - s_0) = \phi(s_0) + (s - s_0) \phi'(s_0) + \frac{(s - s_0)^2}{2!} \phi''(s_0) + \cdots$$

$$\therefore \frac{\phi(s)}{s - s_0} = \frac{\phi(s_0)}{s - s_0} + \phi'(s_0) + \frac{s - s_0}{2!} \phi''(s_0) + \cdots$$

$$\therefore \text{ Residue at } s = s_0 \text{ is } \phi(s_0) = \lim_{s \longrightarrow s_0} (s - s_0) \frac{\phi(s)}{s - s_0}$$

$$= \lim_{s \longrightarrow s_0} (s - s_0) \frac{p(s)}{q(s)}$$

II. Multiple poles.

212

Let $\dfrac{p(s)}{q(s)} = \dfrac{\phi(s)}{(s-s_0)^m}$

$$\phi(s) = \phi(s_0) + (s-s_0)\,\phi'(s_0) + \frac{(s-s_0)^2}{2!}\,\phi''(s_0) + \cdots$$

$$+ \frac{(s-s_0)^{m-1}}{(m-1)!}\,\phi^{(m-1)}(s_0) + \cdots$$

$$\therefore \frac{\phi(s)}{(s-s_0)^m} = \frac{\phi(s_0)}{(s-s_0)^m} + \frac{\phi'(s_0)}{(s-s_0)^{m-1}} + \cdots$$

$$+ \frac{1}{s-s_0}\,\frac{\phi^{(m-1)}(s_0)}{(m-1)!} + \cdots$$

∴ Residue at the pole $s = s_0$ of multiplicity m is given by

$$\frac{1}{(m-1)}\,\phi^{(m-1)}(s_0) = \frac{1}{(m-1)!}\left[\frac{d^{m-1}}{ds^{m-1}}\,\phi(s)\right]_{s=s_0}$$

Example 1

Find $\quad L^{-1}\left\{\dfrac{1}{(s-1)(s-2)(s-3)}\right\}$

Solution:
This is equal to the sum of residues of

$$\frac{e^{st}}{(s-1)\;(s-2)\;(s-3)}$$

at its poles.

s = 1

$$\text{Residue} = \lim_{s \to 1} (s-1) \frac{e^{st}}{(s-1)(s-2)(s-3)}$$

$$= \lim_{s \to 1} \frac{e^{st}}{(s-2)(s-3)} = \frac{e^t}{2}$$

s = 2

$$\text{Residue} = \lim_{s \to 2} \frac{e^{st}}{(s-1)(s-3)} = \frac{e^{2t}}{-1} = -e^{2t}$$

s = 3

$$\text{Residue} = \lim_{s \to 3} \frac{e^{st}}{(s-1)(s-2)} = \tfrac{1}{2} e^{3t}$$

$$\therefore \quad L^{-1} \left\{ \frac{1}{(s-1)(s-2)(s-3)} \right\} = \tfrac{1}{2} e^t - e^{2t} + \tfrac{1}{2} e^{3t}$$

Example 2

$$\text{Find} \quad L^{-1} \left\{ \frac{1}{s^3(s+1)} \right\}$$

Solution:
This is equal to the sum of residues of

$$\frac{e^{st}}{s^2(s+1)}$$

at its poles.
The residue at the double pole at s = 0 is

$$\left\{ \frac{d}{ds} \frac{e^{st}}{s+1} \right\}_{s=0} \left\{ \frac{(s+1) t e^{st} - e^{st}}{(s+1)^2} \right\}_{s=0} = t - 1$$

The residue at s = -1 is

$$\lim_{s \to -1} \frac{e^{st}}{s^2} = e^{-t}$$

$$\therefore \quad L^{-1} \left\{ \frac{1}{s^2(s+1)} \right\} = e^{-t} + t - 1$$

Example 3

214

Find $L^{-1} \left\{ \dfrac{1}{s^2 (s^2 + \omega^2)} \right\}$

Solution:

This is equal to the sum of residues of

$$\frac{e^{st}}{s^2(s-i\omega)(s+i\omega)}$$

at its poles.

Residue at the double pole at $s = 0$ is

$$\left\{ \frac{d}{ds} \frac{e^{st}}{s^2+\omega^2} \right\}_{s=0} = \left\{ \frac{(s^2+\omega^2) t e^{st} - 2s e^{st}}{(s^2+\omega^2)^2} \right\}_{s=0} = \frac{t}{\omega^2}$$

Sum of residues at $s = \pm i\omega$ is

$$\left\{ \frac{e^{st}}{s^2(s+i\omega)} \right\}_{s=i\omega} + \left\{ \frac{e^{st}}{s^2(s-i\omega)} \right\}_{s=-i\omega}$$

$$= -\frac{1}{\omega^3} \left[\frac{e^{i\omega t} - e^{-i\omega t}}{2i} \right] = -\frac{\sin \omega t}{\omega^3}$$

$$\therefore \quad L^{-1} \left\{ \frac{1}{s^2(s^2+\omega^2)} \right\} = \frac{1}{\omega^3} (\omega t - \sin \omega t)$$

2.24. Inversion in the case of branch points

The path of integration of the inversion integral has to be modified if the integrand has a branch point. This is illustrated by the following example.

Suppose it is required to find the inverse transform of $s^{-\frac{1}{2}}$. Here the function $s^{-\frac{1}{2}}$ has a branch point at the origin. We therefore cut the s-plane along the negative x-axis and take the path shown in this sketch.

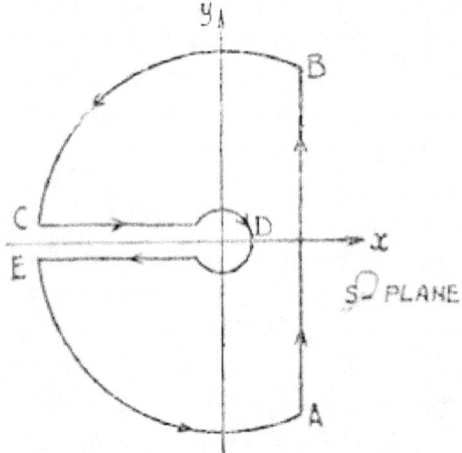

Since $\dfrac{e^{st}}{\sqrt{s}}$ has no singularities within this closed path, hence according to Cauchy's theorem we have:

$$\int_{AB} \frac{e^{st}\,ds}{\sqrt{s}} + \int_{BC+EA} \frac{e^{st}\,ds}{\sqrt{s}} + \int_{CDE} \frac{e^{st}\,ds}{\sqrt{s}} = 0$$

It is easy to show that the integral along the circular path of infinite radius is zero and hence

$$\int_{AB} \frac{e^{st}\,ds}{\sqrt{s}} = - \int_{CDE} \frac{e^{st}\,ds}{\sqrt{s}} = \int_{EDC} \frac{e^{st}\,ds}{\sqrt{s}} \quad \text{in the limit,}$$

when the radius of the circular path is infinite.

Now $\displaystyle\int_{EDC}$ consists of three puts, one around the small circle at D and the others along the straight paths ED and DC. We shall now evaluate each of the three integrals. Consider first the integral around the small circle at D. Taking the radius of this small circle to be equal r, then we can put for points on the circle:

$$s = re^{i\theta}, \quad ds = ire^{i\theta}\,d\theta, \quad s^{\frac{1}{2}} = r^{\frac{1}{2}} e^{i\frac{\theta}{2}}$$

$$\therefore \int_{\text{small circle}} \frac{e^{st}\,ds}{\sqrt{s}} = i \int_{r\to 0} e^{rt(\cos\theta + i\sin\theta)}\, r^{\frac{1}{2}} e^{i\frac{\theta}{2}}\,d\theta = 0$$

Next, consider the integral along ED and put for points on ED

216

$$s = xe^{-i\pi} = -x \, , \, ds = -dx, \, s^{\frac{1}{2}} = x^{\frac{1}{2}} e^{-i\frac{\pi}{2}} = -ix^{\frac{1}{2}}$$

$$\therefore \int_{ED} \frac{e^{st} \, ds}{\sqrt{s}} = -i \int_{\infty}^{0} x^{-\frac{1}{2}} e^{-xt} \, dx = i \int_{0}^{\infty} x^{-\frac{1}{2}} e^{-xt} \, dx$$

Finally along DC, put

$$s = xe^{i\pi} = -x, \, ds = -dx, \, s^{\frac{1}{2}} = x^{\frac{1}{2}} e^{i\frac{\pi}{2}} = ix^{\frac{1}{2}}$$

$$\therefore \int_{DC} \frac{e^{st} \, ds}{\sqrt{s}} = i \int_{0}^{\infty} x^{-\frac{1}{2}} e^{-xt} \, dx$$

Hence the sum of the three integrals is

$$2i \int_{0}^{\infty} x^{-\frac{1}{2}} e^{-xt} \, dx$$

Put $xt = \lambda^2$ where t is constant. $\therefore t \, dx = 2\lambda d\lambda$

$$\therefore 2i \int_{0}^{\infty} x^{-\frac{1}{2}} e^{-xt} dx = 2i \int_{0}^{\infty} \frac{\sqrt{t}}{\lambda} e^{-\lambda^2} \frac{2\lambda}{t} \, d\lambda$$

$$= \frac{4i}{\sqrt{t}} \int_{0}^{\infty} e^{-\lambda^2} \, d\lambda = \frac{4i}{\sqrt{t}} \cdot \frac{\sqrt{\pi}}{2} = 2i \sqrt{\frac{\pi}{t}}$$

Hence,

$$L^{-1} \left\{ \frac{1}{\sqrt{s}} \right\} = \frac{1}{2\pi i} \times 2i \sqrt{\frac{\pi}{t}} = \sqrt{\frac{1}{\pi t}}$$

2.25. Miscellaneous Examples on Laplace Transform

Example 1

1f $L^{-1} \left\{ f(s) \right\} = F(t)$, determine $L^{-1} \left\{ \frac{f(s)}{\sinh cs} \right\}$

$$\frac{f(s)}{\sinh cs} = \frac{2f(s)}{e^{cs} - e^{-cs}} = \frac{2f(s)e^{-cs}}{1 - e^{-2cs}}$$

Now $\dfrac{1}{1-e^{-2cs}} = \displaystyle\sum_{n=0}^{\infty} e^{-2ncs}$

$\therefore \quad \dfrac{f(s)}{\sinh cs} = 2 \displaystyle\sum_{n=0}^{\infty} f(s)e^{(-2nc-c)s}$

$\therefore \quad L^{-1}\left\{ \dfrac{f(s)}{\sinh cs} \right\} = 2 \displaystyle\sum_{n=0}^{\infty} F(t-2nc-c)U(t-2nc-c)$

Example 2

Solve the differential equation

$$x''(t) + x(t) = F(t) \ , \ x(o) = x'(o) = o \quad \text{in which}$$

$$F(t) = 4 \qquad\qquad o \leqslant t \leqslant 2$$
$$= t + 2 \qquad\qquad t > 2$$

Solution:

$$f(s) = \int_{0}^{2} 4e^{-st}\ dt + \int_{2}^{\infty} (t+2)\ e^{-st}\ dt$$

$$= 4\left[\dfrac{e^{-st}}{-s}\right]_{0}^{2} - \dfrac{1}{s}\int_{2}^{\infty}(t+2)\ de^{-st}$$

$$= 4\left[\dfrac{e^{-st}}{-s}\right]_{0}^{2} - \dfrac{1}{s}\left[(t+2)e^{-st} + \dfrac{e^{-st}}{s}\right]_{2}^{\infty}$$

$$= \dfrac{4}{s} + \dfrac{e^{-2s}}{s^2}$$

$$\therefore \quad (s^2+1)\ \bar{x} = \dfrac{4}{s} + \dfrac{e^{-2s}}{s^2}$$

$$\therefore \quad \bar{x} = \dfrac{4}{s(s^2+1)} + \dfrac{e^{-2s}}{s^2(s^2+1)}$$

$$= 4\left[\dfrac{1}{s} - \dfrac{s}{s^2+1}\right] + \left[\dfrac{1}{s^2} - \dfrac{1}{s^2+1}\right]e^{-2s}$$

$$\therefore \quad x(t) = 4-4\cos t + [t-2 - \sin(t-2)]\ U(t-2)$$

Example 3

218

Find $L^{-1} \left\{ \dfrac{1}{\sqrt{s} + 1} \right\}$

Solution:

$$\frac{1}{\sqrt{s}+1} = \frac{\sqrt{s}-1}{s-1} = -\frac{1}{s-1} + \frac{\sqrt{s}}{s-1} = -\frac{1}{s-1} + \frac{s}{\sqrt{s}}\left(\frac{1}{s-1}\right)$$

$$= -\frac{1}{s-1} + \frac{1}{\sqrt{s}}\left(\frac{s-1+1}{s-1}\right) = -\frac{1}{s-1} + \frac{1}{\sqrt{s}}\left(1 + \frac{1}{s-1}\right)$$

$$= -\frac{1}{s-1} + \frac{1}{\sqrt{s}} + \frac{1}{\sqrt{s}\,(s-1)}$$

$$\therefore \quad L^{-1}\left\{\frac{1}{\sqrt{s}+1}\right\} = -e^t + \frac{1}{\sqrt{\pi t}} + e^t\, \mathrm{erf}\,(\sqrt{t})$$

$$= \frac{1}{\sqrt{\pi t}} + e^t\,(\mathrm{erf}\,(\sqrt{t}) - 1)$$

$$= \frac{1}{\sqrt{\pi t}} - e^t\, \mathrm{erfc}\,\sqrt{t}$$

Example 4.

Prove that $\displaystyle\int_{-\infty}^{\infty} \frac{x \sin tx}{a^2 + x^2}\, dy = \pi e^{-at} \quad (a > 0,\ t > 0)$.

By differentiating under the sign of integration w.r.t. a

deduce the value of $\displaystyle\int_{-\infty}^{\infty} \frac{x \sin x}{(a^2 + x^2)^2}\, dx$

Solution:

$$\int_{-\infty}^{\infty} \frac{x \sin tx}{a^2 + x^2}\, dx = 2\int_{0}^{\infty} \frac{x \sin tx}{a^2 + x^2}\, dx, \quad \text{the integrand being}$$

an even function of x.

$$L\left\{\int_{0}^{\infty} \frac{x \sin tx}{a^2 + x^2}\, dx\right\} = \int_{0}^{\infty} e^{-st}\left\{\int_{0}^{\infty} \frac{x \sin tx}{a^2 + x^2}\, dx\right\} dt$$

219

$$= \int_0^\infty \frac{x}{a^2 + x^3} \left\{ \int_0^\infty e^{-st} \sin tx \, dt \right\} dx$$

$$= \int_0^\infty \frac{x}{a^2 + x^2} \cdot \frac{x}{s^2 + x^2} \, dx = \int_0^\infty \frac{x^2 \, dx}{(x^2 + a^2)(x^2 + s^2)}$$

$$= \frac{1}{s^2 - a^2} \int_0^\infty \left[\frac{s^2}{x^2 + s^2} - \frac{a^2}{x^2 + a^2} \right] dx$$

$$= \frac{1}{s^2 - a^2} \left[s \tan^{-1} \frac{x}{s} - a \tan^{-1} \frac{x}{a} \right]_0^\infty$$

$$= \frac{1}{s^2 - a^2} \cdot \frac{\pi}{2} (s - a) = \frac{\pi}{2(s+a)}$$

$$\therefore \int_0^\infty \frac{x \sin tx}{a^3 + x^3} \, dx = \frac{\pi}{2} e^{-at} \, , \quad \therefore \int_{-\infty}^\infty \frac{x \sin tx}{a^2 + x^2} \, dx = \pi e^{-at}$$

Putting t =1 and differentiating under the sign of integration w.r,t a we get

$$- 2a \int_{-\infty}^\infty \frac{x \sin x}{(a^2 + x^2)^2} \, dx = - \pi e^{-a}$$

$$\therefore \int_{-\infty}^\infty \frac{x \sin x}{(a^2 + x^2)^2} \, dx = \frac{1}{2a} \pi e^{-a}$$

Example 5

Solve, using an inversion integral, the differential equation

$$(D^2 - 2D + 2)(D^2 + 2D - 3) x = 0, t > 0$$

given that $x(o) = 1$, $x'(o) = o$, $x''(o) = 6$, $x'''(o) = - 14$

Solution:

The equation may be written

$$(D^4 - 5D^2 + 10D - 6) \, x = 0$$

$$s^4 \bar{x} - s^3 x(o) - s^2 x'(o) - s x''(o) - x'''(o)$$

$$-5 \, [s^2 \bar{x} - s x(o) - x'(o)] + 10 \, [s \bar{x} - x(o)] - 6\bar{x} = o$$

$$\therefore \quad [s^4 - 5s^2 + 10s - 6] \, \bar{x} = s^3 + 6s - 14 - 5s + 10$$

i.e $\quad (s^2 - 2s + 2) (s^2 + 2s - 3) \, \bar{x} = s^3 + s - 4$

$$\therefore \quad \bar{x} = \frac{s^3 + s - 4}{(s^2 - 2s + 2) \, (s + 3) \, (s - 1)}$$

$\therefore \quad x = $ sum of residues of $\dfrac{(s^3 + s - 4) \, e^{st}}{(s^2 - 2s + 2)(s + 3)(s - 1)}$ at its poles.

Residue at $\quad s = 1 \pm i$ is $\frac{1}{2} (1 \mp i) \, e^{(1 \pm i)t}$

Residue at $\quad s = -3$ is $\frac{1}{2} \, e^{-3t}$

Residue at $\quad s = 1$ is $-\frac{1}{2} \, e^t$

Adding we get

$$x = e^t \, (\cos t + \sin t) - \frac{1}{2} \, e^t + \frac{1}{2} e^{-3t}$$

Example 6

Prove that $\displaystyle\int_0^\infty J_0(t) \, dt = 1$

We have shown that

$$L \left\{ J_0(t) \right\} = \frac{1}{\sqrt{s^2 + 1}} \quad (2.10)$$

$$\therefore \quad \int_0^\infty e^{-st} J_0(t) \, dt = \frac{1}{\sqrt{s^2 + 1}}$$

Let s $\rightarrow$ 0 + we get

$$\int_0^\infty J_o(t)\ dt = 1$$

Example 7

Find $L\left\{ t\ \sinh\ 3t\ \right\}$

$$L\left\{ t\ \sinh\ 3t\ \right\} = -\frac{d}{ds}\left\{ \frac{3}{s^2-9} \right\} = \frac{6s}{(s^2-9)^2}$$

Example 8

Evaluate $\displaystyle\int_0^\infty te^{-2t}\ \sin\ t\ dt$

Solution:

$$L\left\{ t\ \sin\ t \right\} = -\frac{d}{ds}\ \frac{1}{s^2+1} = \frac{2s}{(s^2+1)^2}$$

$$\therefore \int_0^\infty e^{-st}\ t\ \sin\ t\ dt = \frac{2s}{(s^2+1)^2}$$

Let $s \longrightarrow 2$ $\therefore \displaystyle\int_0^\infty te^{-2t}\sin\ t\ dt = \frac{4}{25}$

Example 9

Find $L\left\{ F(t) \right\}$ where

$$F(t) = t \qquad 0 < t < 1$$

$$= 0 \qquad 1 < t < 2$$

$$F(t+2) = F(t)$$

$$L\left\{F(t)\right\} = \frac{\displaystyle\int_0^2 e^{-st} F(t)\, dt}{1 - e^{-2s}} = \frac{\displaystyle\int_0^1 e^{-st} t\, dt}{1 - e^{-2s}}$$

$$\int_0^1 e^{-st} t\, dt = -\frac{1}{s}\int_0^1 t\, de^{-st} = -\frac{1}{s}\left[te^{-st} + \frac{e^{-st}}{2} \right]_0^1$$

$$= \frac{1}{s}\left[\frac{1}{s} - e^{-s} - \frac{e^{-s}}{s} \right] = \frac{1}{s^2}\left[1 - (s+1)\, e^{-s} \right]$$

$$\therefore\ L\left\{F(t)\right\} = \frac{1}{s^2}\ \frac{1-(s+1)e^{-s}}{1-e^{-2s}}$$

Example 10

Express in terms of the unit step function

$$\begin{aligned}
F(t) &= \cos 2t && 0 < t < \pi,\\
&= \cos 4t && \pi < t < 2\pi,\\
&= \cos 6t && t > 2\pi
\end{aligned}$$

It follows directly from the definition of the unit step function that

$$F(t) = \cos 2t + (\cos 4t - \cos 2t)\, U(t - \pi)$$
$$+ (\cos 6t - \cos 4t)\, U(t - 2\pi)$$

Example 11

Evaluate $\displaystyle\int_0^\infty \frac{e^{-t} \sin t}{t}\, dt$

223

$$L \left\{ \sin t \right\} = \frac{1}{s^2 + 1}$$

$$L \left\{ \frac{\sin t}{t} \right\} = \int_s^\infty \frac{dx}{x^2 + 1} = \left[\tan^{-1} x \right]_s^\infty = \frac{\pi}{2} - \tan^{-1} s$$

$$\therefore \int_0^\infty \frac{e^{-st} \sin t}{t} \, dt = \frac{\pi}{2} - \tan^{-1} s$$

Let s → 1

$$\therefore \int_0^\infty \frac{e^{-t} \sin t}{t} \, dt = \frac{\pi}{2} - \tan^{-1} 1 = \frac{\pi}{4}$$

Example 12

$$\text{Evaluate} \int_0^t J_0(u) \, J_1(t-u) \, du$$

Solution:

$$\int_0^t J_0(u) \, J_1(t-u) \, du = J_0(t)^* \, J_1(t)$$

$$L \left\{ \int_0^t J_0(u) \, J_1(t-u) \, du \right\} = \frac{1}{\sqrt{s^2+1}} \left(1 - \frac{s}{\sqrt{s^2+1}} \right)$$

$$= \frac{1}{\sqrt{s^2+1}} - \frac{s}{s^2+1}$$

$$\therefore \int_0^t J_0(u) \, J_1(t-u) \, du = J_0(t) - \cos t$$

Example 13

Solve the integral equation $\int_0^t Y(t-\lambda) \ Y(\lambda) \ d\lambda = 8 \sin 2t$

$$Y(t)^* \ Y(t) = 8 \sin 2t$$

$$\therefore \ [y(s)]^2 = \frac{16}{s^2+4} \qquad \therefore \ y(s) = \pm \frac{4}{\sqrt{s^2+4}}$$

Now $L \left\{ J_0(at) \right\} = \dfrac{1}{\sqrt{s^2+a^2}}$

$$\therefore \ Y(t) = \pm \ 4J_0(2t)$$

Example 14

Find L { Log t }
1

we have shown in 1.2 that $L \left\{ t^k \right\} = \dfrac{\Gamma(k+1)}{s^{k+1}}$ where $k > -1$

i.e $\int_0^\infty e^{-st} t^k \ dt = \dfrac{\Gamma(k+1)}{s^{k+1}}$

Differentiate w,r.t.. k

$$\int_0^\infty e^{-st} t^k \ \log t \ dt = \frac{\Gamma'(k+1) - \Gamma(k+1) \ \log \ s}{s^{k+1}}$$

Put $k = 0$

$$\int_0^\infty e^{-st} \log t \ dt = \frac{\Gamma'(1) - \Gamma(1) \ \log s}{s}$$

$$\therefore \ L \left\{ \log t \right\} = \frac{\Gamma'(1) - \log s}{s}$$

Example 15
Solve the integral equation

225

$$Y(t) = a \sin 8t + 6 \int_0^t Y(\lambda) \sin 8 (t-\lambda) \, d\lambda$$

$$\therefore \quad Y(t) = a \sin 8t + 6Y(t)' \sin 8t$$

$$y(s) = \frac{8a}{s^2+64} + 6y(s) \frac{8}{s^2+64}$$

$$\therefore \quad y(s) \left[1 - \frac{48}{s^2+64} \right] = \frac{8a}{s^2+64}$$

$$\text{i.e} \quad y(s) \cdot \frac{s^2+16}{s^2+64} = \frac{8a}{s^2+64}$$

$$\therefore \quad y(s) = \frac{8a}{s^2+16} \quad \therefore \quad Y(t) = 2a \sin 4t$$

Example 16

Evaluate $\displaystyle\int_0^\infty \int_0^t \frac{e^{-t} \sin u}{u} \, dt \, du$

$$\text{Integral} = \int_0^\infty e^{-t} \left\{ \int_0^t \frac{\sin u}{u} \, du \right\} dt$$

$$L\left\{ \sin t \right\} = \frac{1}{s^2+1}$$

$$L\left\{ \frac{\sin t}{t} \right\} = \int_s^\infty \frac{dx}{x^2+1} = \left[\tan^{-1}x \right]_s^\infty = \frac{\pi}{2} - \tan^{-1}s = \tan^{-1}\frac{1}{s}$$

$$L\left\{ \int_0^t \frac{\sin u}{u} \, du \right\} = \frac{1}{s} \tan^{-1} \frac{1}{s}$$

$$\text{i.e} \quad \int_0^\infty e^{-st} \left\{ \int_0^t \frac{\sin u}{u} \, du \right\} dt = \frac{1}{s} \tan^{-1} \frac{1}{s}$$

Let s → 1

$$\int_0^\infty e^{-t} \left\{ \int_0^t \frac{\sin u}{u} \, du \right\} dt = \tan^{-1} 1 = \frac{\pi}{4}$$

$$\therefore \int_0^\infty \int_0^t e^{-t} \frac{\sin u}{u} \, dt \, du = \frac{\pi}{4}$$

Example 17

Evaluate $\displaystyle\int_0^\infty \frac{e^{-2t} - e^{-4t}}{t} \, dt$

$$L \left\{ e^{-2t} \right\} = \frac{1}{s+2}, \quad L \left\{ e^{-4t} \right\} = \frac{1}{s+4}$$

$$\therefore \; L \left\{ e^{-2t} - e^{-4t} \right\} = \left[\frac{1}{s+2} - \frac{1}{s+4} \right]$$

$$\therefore \; L \left\{ \frac{e^{-2t} - e^{-4t}}{t} \right\} = \int_s^\infty \left[\frac{1}{x+2} - \frac{1}{x+4} \right] dx$$

$$= \left[\log \frac{x+2}{x+4} \right]_s^\infty = \log \frac{s+4}{s+2}$$

i.e $\displaystyle\int_0^\infty e^{-st} \frac{e^{-2t} - e^{-4t}}{t} \, dt = \log \frac{s+4}{s+2}$

Let $s \rightarrow 0$ $\therefore \displaystyle\int_0^\infty \frac{e^{-2t} - e^{-4t}}{t} \, dt = \log \frac{4}{2} = \log 2$

Example 18

Evaluate $\displaystyle\int_0^\infty e^{-t} J_0(t) \, dt$

227

We have seen that $L\left\{J_0(t)\right\} = \dfrac{1}{\sqrt{s^2+1}}$

$$\therefore \int_0^\infty e^{-st} J_0(t)\ dt = \dfrac{1}{\sqrt{s^2+1}}$$

Let $s \longrightarrow 1$ $\therefore \displaystyle\int_0^\infty e^{-t} J_0(t)\ dt = \dfrac{1}{\sqrt{2}}$

Example 19

Evaluate $L^{-1}\left\{\dfrac{1}{s^2(s^2+1)}\right\}$ (i) by resolving into partial fractions, (ii) by convolution, (iii) from $L^{-1}\left\{\dfrac{1}{s^2+1}\right\}$, (iv) by using an inversion integral.

(i) $L^{-1}\left\{\dfrac{1}{s^2(s^2+1)}\right\} = L^{-1}\left\{\dfrac{1}{s^2} - \dfrac{1}{s^2+1}\right\} = t - \sin t$

(ii) $L^{-1}\left\{\dfrac{1}{s^2} \cdot \dfrac{1}{s^2+1}\right\} = t^* \sin t = \displaystyle\int_0^t (t-\lambda)\sin\lambda\ d\lambda$

$= -\displaystyle\int_0^t (t-\lambda)\ d\cos\lambda = \left[-(t-\lambda)\cos\lambda - \sin\lambda\right]_0^t = -\sin t + t$

(iii) $L^{-1}\left\{\dfrac{1}{s^2+1}\right\} = \sin t$ $\therefore$ $L^{-1}\left\{\dfrac{1}{s(s^2+1)}\right\}$

$= \displaystyle\int_0^t \sin\tau\ d\tau = \left[-\cos\tau\right]_0^t = 1 - \cos t$

$L^{-1}\left\{\dfrac{1}{s^2(s^2+1)}\right\} = \displaystyle\int_0^t (1-\cos\tau)\ d\tau = \left[\tau-\sin\tau\right]_0^t = t - \sin t$

228

(iv) $L^{-1} \left\{ \dfrac{1}{s^2(s^2+1)} \right\}$ = sum of residues of $\left\{ \dfrac{e^{st}}{s^2(s-i)(s+i)} \right\}$ at its poles.

$$s = o \quad \text{residue} = \frac{1}{1!} \left[\frac{d}{ds} \left(\frac{e^{st}}{s^2+1} \right) \right]_{s=0}$$

$$= \left[\frac{te^{st}(s^2+1) - 2s\,e^{st}}{(s^2+1)^2} \right]_{s=0} = t$$

Sum of residues at $s = \pm i$ is $\left[\dfrac{e^{st}}{s^2(s+i)} \right]_{s=i} + \left[\dfrac{e^{st}}{s^2(s-i)} \right]_{s=-i}$

$$= \frac{e^{it}}{-2i} + \frac{e^{-it}}{2i} = \sin t$$

$$\therefore \quad L^{-1} \left\{ \frac{1}{s^2(s^2+1)} \right\} = t - \sin t$$

Example 20

Solve the equation

$$tY''(t) + (t-1)\,Y'(t) - Y(t) = o, \text{ given that } Y(o)=5,\ Y(\infty)=o$$

$$- \frac{d}{ds}[s^2 y(s) - sY(o) - Y'(o)] - \frac{d}{ds}[sy(s) - Y(o)]$$

$$- [sy(s) - Y(o)] - y(s) = o$$

$$\therefore \quad - \frac{d}{ds}[s^2 y(s) - 5s - A] - \frac{d}{ds}[sy(s) - 5] - [sy(s) - 5] - y(s) = o$$

$$\therefore \quad - s^2 \frac{dy}{ds} - 2sy + 5 - s\frac{dy}{ds} - y - sy + 5 - y = o$$

$$- (s^2 + s)\frac{dy}{ds} - 3sy - 2y + 10 = 0$$

$$\therefore \quad \frac{dy}{ds} + \frac{3s+2}{s^2+s}\, y = \frac{10}{s^2+s}$$

which is a linear differential equation of the first order.

229

Now $\quad \dfrac{3s+2}{s^2+s} = \dfrac{2}{s} + \dfrac{1}{s+1}$

Integrating factor $= e^{\int \left(\frac{2}{s} + \frac{1}{s+1}\right) ds} = e^{2\log s + \log(s+1)}$

$$= e^{\log s^2(s+1)} = s^2(s+1)$$

$$\therefore \quad y\, s^2(s+1) = \int \dfrac{10}{s(s+1)}\, s^2(s+1)\ ds + C$$

$$= 10 \int s\, ds + C = 5s^2 + C$$

$$\therefore \quad y(s) = \dfrac{5}{s+1} + \dfrac{C}{s^2(s+1)}$$

$$= \dfrac{5}{s+1} + C\left(\dfrac{1}{s^2} - \dfrac{1}{s} + \dfrac{1}{s+1}\right)$$

$$\therefore \quad Y(t) = 5e^{-t} + C(\,t - 1 + e^{-t}\,)$$

and since $Y(t) \longrightarrow o$ when $t \longrightarrow \infty$, then $C = o$

and hence $Y(t) = 5e^{-t}$

Example 21

Solve the differential equation

$$\dfrac{d^2 y}{dx^2} - a^2 y = \delta(x-b)\,,\ o<x<1,\ o<b<1\,,$$

with $y = 0$ when $x = 0$ and $x = 1$

$$s^2 \overline{y} - sy(o) - y'(o) - a^2 \overline{y} = e^{-bs}$$

Put $y'(o) = A \qquad \therefore\ (s^2 - a^2)\,\overline{y} = A + e^{-bs}$

$$\overline{y} = \dfrac{A}{s^2 - a^2} + \dfrac{e^{-bs}}{s^2 - a^2}$$

230

$$y = \frac{A}{a} \sinh ax + \frac{1}{a} \sinh a(x-b) \ U(x-b)$$

Since $y=0$ when $x=1$ we get

$$o = \frac{A}{a} \sinh al + \frac{1}{a} \sinh a(1-b)$$

$$\therefore \quad A = -\frac{\sinh a(1-b)}{\sinh al}$$

$$y = -\frac{\sinh a(1-b) \sinh ax}{a \sinh al} + \frac{1}{a} \sinh a(x-b) \ U(x-b)$$

i.e $\quad y = -\frac{\sinh a(1-b) \sinh ax}{a \sinh al} \qquad o < x < b$

$$= -\frac{\sinh a(1-b) \sinh ax}{a \sinh al} + \frac{1}{a} \sinh a(x-b)$$

$$= -\frac{\sinh ab \sinh a(1-x)}{a \sinh al} \qquad b < x < 1$$

2.26. Exercises on Laplace Transforms

(1) Find (i) $L^{-1} \left\{ \dfrac{5e^{-3s}}{s} - \dfrac{e^{-s}}{s} \right\}$, (ii) $L^{-1} \left\{ \dfrac{e^{-4s}}{(s+2)^3} \right\}$

[Ans. (i) $5U(t-3) - U(t-1)$, (ii) $\frac{1}{2}(t-4)^2 e^{-2(t-4)} U(t-4)$]

(2) If $F(t)$ is to be continuous for $t \geqslant o$ and

$$F(t) = L^{-1} \left\{ \frac{(1-e^{-2s})(1-3e^{-2s})}{s^2} \right\}$$

evaluate $F(1)$, $F(3)$, $F(5)$.

[Ans. $1, -1, -4$]

231

(3) Find the Laplace transform of

$$F(t) = E \sin \omega t \qquad o < t < \frac{\pi}{\omega}$$

$$= o \qquad\qquad t > \frac{\pi}{\omega}$$

$$\left[\text{Ans. } f(s) = \frac{E\omega}{s^2 + \omega^2} \left(1 + e^{-\frac{s\pi}{\omega}} \right) \right]$$

(4) Find $L\left\{ t^2 \sin kt \right\}$, $L\left\{ t^2 \cos kt \right\}$

$$\left[\text{Ans. } \frac{2k(3s^2 - k^2)}{(s^2 + k^2)^3} \quad s > o ; \quad \frac{2s(s^2 - 3k^2)}{(s^2 + k^2)^3}, \quad s > o \right]$$

(5) Find the Laplace transform of the periodic function $\psi(t,c)$ where

$$\psi(t,c) = 1 \qquad o < t < c \quad ,$$

$$= o \qquad c < t < 2c \quad ,$$

$$\psi(t + 2c) = \psi(t,c)$$

$$\left[\text{Ans. } \frac{1}{s(1 + e^{-cs})} \right]$$

(6) Evaluate $L\left\{ \frac{\sin kt}{t} \right\}$ $\left[\text{Ans. } \tan^{-1} \frac{k}{s} , \ s > o \right]$

(7) Evaluate $L\left\{ \frac{\sinh kt}{t} \right\}$ $\left[\text{Ans. } \tfrac{1}{2} \log \frac{s + k}{s - k}, \quad s > k > o \right]$

(8) Evaluate $L\left\{ \frac{1 - \cosh kt}{t} \right\}$ $\left[\text{Ans. } \tfrac{1}{2} \log \left(1 - \frac{k^2}{s^2} \right), \ s > k > o \right]$

(9) Evaluate $F(t) = L^{-1}\left\{ \frac{1}{s^3 (1 - e^{-2s})} \right\}$ and compute F(5).

$$\left[\text{Ans. } F(t) = \tfrac{1}{2} \sum_{n=o}^{\infty} (t - 2n)^2 \ U(t - 2n), \quad F(5) = 17.5 \right]$$

(10) If $\Phi(t) = L^{-1}\left\{ \frac{3}{s^4 \sinh 3s} \right\}$, compute $\Phi(10)$ [Ans 344]

(11) Prove that $L^{-1}\left\{f(s)\ \tanh\ cs\right\} = F(t)$

$$+ 2 \sum_{n=1}^{\infty} (-1)^{n}\ F(t-2nc)\ U(t-2nc)\quad c>0,s>0.$$

(12) Prove that

$$L^{-1}\left\{\frac{f(s)}{\cosh\ cs}\right\} = 2 \sum_{n=0}^{\infty} (-1)^{n}$$

$$F(t-2n\ c-c)\ U(t-2nc-c),\quad c>0,\ s>0.$$

(13) Compute $x(1)$ and $x(4)$ for the function $x(t)$ which satisfies the boundary-value problem

$$x''(t) + 2x'(t) + x(t) = 2 + (t-3)\ U(t-3)$$

given that $x(o) = 2$ and $x'(o) = 1$

$$\left[\text{Ans. } x(1) = 2 + \frac{1}{e}\ ,\quad x(4) = 1 + \frac{3}{e} + \frac{4}{e^4}\right]$$

(14) Solve the differential equation

$$x''(t) + 2x'(t) + x(t) = F(t);\quad x(o) = x'(o) = o$$

$$\left[\text{Ans. } x(t) = \int_{0}^{t} \lambda\ e^{-\lambda}\ F(t-\lambda)\ d\lambda\right]$$

(15) Solve the differential equation

$$y''(t) + 4y'(t) + 13\ y(t) = F(t)\ ;\ y(o) = o,\ y'(o) = o.$$

$$\left[\text{Ans. } y(t) = \frac{1}{3} \int_{0}^{t} e^{-2\lambda}\ \sin 3\lambda\ F(t-\lambda)\ d\lambda\right]$$

Solve the differential equations

(16)

$$x''(t) + 6x'(t) + 9x(t) = F(t); \quad x(o) = A, \ x'(o) = B.$$

$$\left[\text{Ans.} \quad x(t) = e^{-3t} \left[A + (B + 3A) t \right] \right.$$

$$\left. + \int_{o}^{t} \lambda e^{-3\lambda} F(t-\lambda) \, d\lambda \right]$$

(17)

$$F(t) = 1 + 2 \int_{o}^{t} F(t-\lambda) \, e^{-2\lambda} \, d\lambda$$

[Ans. $F(t) = 1 + 2t$]

(18)

$$F(t) = 4t^2 - \int_{o}^{t} F(t-\lambda) \, e^{-\lambda} \, d\lambda$$

[Ans. $F(t) = -1 + 2t + 2t^2 + e^{-2t}$]

(19)

$$F(t) = 8t^2 - 3 \int_{0}^{t} F(\lambda) \sin(t-\lambda) \, d\lambda$$

[Ans. $F(t) = 2t^3 + 3 - 3 \cos 2t$]

Solve the following differential equations and check by using an inversion integral:

(20) $(D^3+1) x = 1$, $x(o) = x'(o) = x''(o) = o$

[Ans. $x = 1 - \dfrac{1}{3} e^{-t} - \dfrac{2}{3} e^{\frac{1}{2}t} \cos \frac{1}{2} t \sqrt{3}$]

(21) $(D+1)(D+2)(D+3) x = 1 + t + t^2$

given that $x(o) = x'(o) = x''(o) = o$

[Ans. $\dfrac{35}{54} - \dfrac{4}{9} t + \dfrac{1}{6} t^2 - e^{-t} + \frac{1}{2} e^{-2t} - \dfrac{4}{27} e^{-3t}$]

(22) $D(D-1)x = t^2$ where $x(o) = x_0$ and $x'(o) = x_1$

[Ans, $(x_0 - x_1) + x_1 e^t - 2 \left(1 + t + \dfrac{t^2}{2!} + \dfrac{t^3}{3!} - e^t\right)$]

(23) $(D^2 - 3D + 2) x = e^t$ where $x(o) = x_0$ and $x'(o) = x_1$

[Ans. $(x_1 - x_0 + 1) e^{2t} + (2x_0 - x_1 - 1 - t) e^t$]

(24) Indicate which of the following are null functions

(i) $F(t) = 2, \; t = 7$.(ii) $F(t) = 4 \quad 1 \leqslant t \leqslant 3$

$= 0$ otherwise

[Ans. (i) is a null function but (ii) is not.]

(25) Given that $L \left\{ \sin \sqrt{t} \right\} = \dfrac{\sqrt{\pi}}{2s^{3/2}} e^{-\frac{1}{4s}}$, show that

$$ L \left\{ \dfrac{\cos \sqrt{t}}{\sqrt{t}} \right\} = \dfrac{\sqrt{\pi}}{s^{1/2}} e^{-\frac{1}{4s}} $$

(26) If $L \left\{ F(t) \right\} = \dfrac{e^{-1/s}}{s}$ Find $L \left\{ e^{-t} F(4t) \right\}$

$$ \left[\text{Ans.} \quad \dfrac{e^{-\frac{4}{s+1}}}{s+1} \right] $$

(27) Show that $L \left\{ \displaystyle\int_0^t \dfrac{1 - e^{-\tau}}{\tau} \, d\tau \right\} = \dfrac{1}{s} \log \left(1 + \dfrac{1}{s}\right)$

(28) Find $L \{ F(t)\}$ where

$F(t) = t^2 \quad o < t < 2,$

$F(t + 2) = F(t)$

$$ \left[\text{Ans.} \quad \dfrac{2 - 2e^{-2s} - 4se^{-2s} - 4s^2 e^{-2s}}{s^3(1 - e^{-2s})} \right] $$

(29) Express in terms of the unit step function

$$F(t) = t^3 \qquad o < t < 3$$
$$= 5t^2 \qquad t > 3$$

[Ans. $F(t) = t^3 + (5t^2 - t^3) U(t-3)$]

(30) Prove that $\int_0^\infty t^3 e^{-t} \sin t \, dt = o$

(31) Find $L \left\{ F(t) \right\}$ where

$$F(t) = \cos t \qquad o < t < \pi$$
$$= \sin t \qquad t > \pi$$

$\left[\text{ Ans. } \dfrac{s + (s-1) e^{-\pi s}}{s^2 + 1} \right]$

(32) Prove that $L \left\{ \sin^5 t \right\} = \dfrac{120}{(s^2+1)(s^2+9)(s^2+25)}$

(33) Prove that $\int_0^\infty \cos x^2 \, dx = \tfrac{1}{2} \sqrt{\dfrac{\pi}{2}}$

(37) Find by convolution

i) $L^{-1} \left\{ \dfrac{1}{(s+2)^2 \, (s-2)} \right\}$

$\left[\text{ Ans. } \dfrac{1}{16} (e^{2t} - e^{-2t} - 4t \, e^{-2t}) \right]$

ii) $L^{-1} \left\{ \dfrac{s^2}{(s^2+4)^2} \right\}$ $\left[\text{Ans. } \dfrac{1}{2} t \cos 2t + \dfrac{1}{4} \sin 2t \right]$

iii) $L^{-1} \left\{ \dfrac{1}{(s+1) \, (s^2+1)} \right\}$

$\left[\text{ Ans. } \dfrac{1}{2} (\sin t - \cos t + e^{-t}) \right]$

(35) Using Heaviside's expansion formula, find

$$L^{-1} \left\{ \dfrac{2s + 5}{(s+2) \, (s+3)} \right\}$$

[Ans. $e^{-2t} + e^{-3t}$]

236

(36) Prove that $L^{-1} \left\{ \dfrac{1}{s} \cos \dfrac{1}{s} \right\} = 1 - \dfrac{t^2}{(2!)^2} + \dfrac{t^4}{(4!)^2} - \dfrac{t^6}{(6!)^2} + \cdots$

(37) Solve the equation

$$tY''(t) + 2Y'(t) + tY(t) = 0$$

given that $Y(o+) = 1$ and $y(\pi) = 0$ $\left[\text{Ans. } Y(t) = \dfrac{\sin t}{t} \right]$

(38) Solve the equation $Y''(t) + tY'(t) - Y(t) = 0$,

given that $Y(o) = 0$, $Y'(o) = 1$ $[\text{Ans. } Y(t) = t]$

(39) Solve the equation $tY''(t) + (1-2t)\, Y'(t) - 2Y(t) = 0$,

given that $Y(o) = 1$, $Y'(o) = 2$. $[\text{Ans. } Y(t) = e^{2t}]$

(40) Solve the integral equation

$$Y(t) = t^2 + \int_o^t Y(\lambda) \sin (t-\lambda)\ d\lambda$$

$$\left[\text{Ans } Y(t) = t^2 + \dfrac{1}{12} t^4 \right]$$

(41) Solve for $Y(t)$, the second order difference equation

$Y(t) - (a+b)\, Y(t-h) + ab\, Y(t-2h) = F(t)$, $h > o$

where $Y(t) = o$ and $F(t) = o$ when $t < o$ in the case when
(i) $a \neq b$, (ii) $a = b$

$\Big[\text{Ans. i)}$

$$Y(t) = \dfrac{1}{a-b} \sum_{n=o}^{\infty} (a^{n+1} - b^{n+1})\ F(t-nh)\ U(t-nh)$$

$$\text{ii) } Y(t) = \sum_{n=o}^{\infty} (n+1)\, a^n\ F(t-nh)\ U(t-nh) \Big]$$

(42) Solve the integral equation

$$Y(t) = 6t + \int_0^t Y(t-\lambda) \sin \lambda \; d\lambda$$

[Ans, $Y(t) = t^3 + 6t$]

Solve the equations

(43)

$$Y''(t) + 2tY'(t) - 4Y(t) = 1, \; Y(o) = Y'(o) = o$$

$$\left[\text{Ans.} \; Y(t) = \frac{t^2}{2} \right]$$

(44)

$$tY''(t) - (1+t) Y'(t) + 2Y(t) = t-1$$

given that $Y(o) = o$ [Ans. $Y(t) = t + At^2$]

(45)

$$Y'(t) + k^2 \int_0^t Y(\lambda) \cosh k \; (t-\lambda \; d\lambda = o$$

$$\left[\text{Ans.} \; Y(t) = C \left(t - \frac{k^2 t^2}{2} \right) \right]$$

(46)

Prove that $L \left\{ \dfrac{e^t - \cos t}{t} \right\} = \dfrac{1}{2} \log \dfrac{s^2 + 1}{(s-1)^2}$

(47) Prove that $\displaystyle \int_0^\infty \frac{\cos t \; x}{x^2 + a^2} \; dx = \frac{\pi}{2a} e^{-at}$, $a > o, \; t \geqslant o$

238

(48) Prove that $\displaystyle\int_0^\infty \frac{1 - \cos 2tx}{x^2} dx = \pi t$

(49) Find using an inversion integral $L^{-1}\left\{\dfrac{2s + 1}{s(s^2+1)}\right\}$

[Ans. $1 - \cos t + 2 \sin t$]

(50) Prove that $L\left\{t J_o(at)\right\} = \dfrac{s}{(s^2 + a^2)^{3/2}}$

(51) Find $L\left\{t J_1(t)\right\}$ $\left[\text{Ans. } \dfrac{1}{(s^2 + 1)^{3/2}} \right]$

(52) Solve the integral equation

$$Y(t) = \cos t - 3\int_0^t Y(\lambda) \sin (t-\lambda) d\lambda$$

[Ans. $Y(t) = \cos 2t$]

(53) Solve the difference equations

(i) $Y(t) - Y(t-2) = t$ where $Y(t) = o$ when $t < o$

(ii) $Y'(t) - Y(t-1) = t$ where $Y(t) = o$ when $t \leq o$

[Ans (i) $Y(t) = t + (t-2)U(t-2) + (t-4)U(t-4) + \cdots$

(ii) $Y(t) = \dfrac{t^2}{2!} + \dfrac{(t-1)^3}{3!}U(t-1) + \dfrac{(t-2)^4}{4!} U(t-2) + \cdots$]

(54) Find the Laplace transform of the wave F(c,t) defined by

239

$$F(c,t) = 0 \qquad 0 < t < c$$
$$= t\text{-}c \quad c < t < 2c$$

$$F(t+2c) = F(t)$$

$$\left[\text{Ans.}\quad \frac{e^{-cs} - (cs+1)\ e^{-2cs}}{s^2(1-e^{-2cs})}\right]$$

(55) Find $L^{-1}\left\{\dfrac{1}{s^2(s-1)}\right\}$ (i) by partial fractions

(ii) by convolution   (iii) from $L^{-1}\left\{\dfrac{1}{s-1}\right\}$

(iv) by an inversion integral [Ans. $e^t - (t+1)$]

(56) Solve $Y(t) + 2\displaystyle\int_0^t Y(\lambda)\ \cosh(t-\lambda)d\lambda = \sinh t$

$$\left[\text{Ans. } Y(t) = \frac{1}{\sqrt 2}\ e^{-t}\ \sinh \sqrt 2\, t\right]$$

(57) Find $L^{-1}\left\{\dfrac{1}{s(s^2-1)}\right\}$ using an inversion integral.

[Ans. $-1 + \tfrac{1}{2}\ e^t + \tfrac{1}{2}\ e^{-t}$]

(58) Solve $Y'(t) = \sin t + \displaystyle\int_0^t Y(t-\lambda)\ \cos \lambda\ d\lambda$, given that

$Y(0) = 0$

$$\left[\text{Ans. } Y(t) = \frac{t^2}{2}\right]$$

(59) Find the Laplace transform of F(c,t) defined by

F(c,t) = 1 $o < t < c$

= 0 $c < t < 2c$

= 1 $2c < t < 3c$

F(c,t+3c) = F(c,t)

$$\left[\text{Ans.} \quad L\left\{ F(c,t) \right\} = \frac{1-e^{-cs} + e^{-2cs} - e^{-3cs}}{s(1-e^{-3cs})} \right]$$

(60) Solve $Y'(t) = \sinh t - \int_0^t Y(t-\lambda) \cosh \lambda \, d\lambda$, $Y(o) = o$

$$[\text{Ans} \quad Y(t) = \tfrac{1}{2} t^2]$$

(61) Solve the difference equation

$$Y(t) - 4Y(t-h) + 4Y(t-2h) = t^2$$

$$[\text{Ans.} \quad Y(t) = t^2 + 4(t-h)^2 U(t-h) + 12(t-2h)^2 U(t-2h) + ..]$$

(62) Find the Laplace transform of the wave F(c,t) defined by

F(c,t) =   2 $o < t < c$

= -2 $c < t < 2c$

=   2 $2c < t < 3c$

F(c,t+3c) = F(c,t)

$$\left[\text{Ans.} \quad \frac{2[1-2e^{-cs} + 2e^{-2cs} - e^{-3cs}]}{s(1-e^{-3cs})} \right]$$

(63) Solve $Y'(t) + \int_0^t Y(\lambda) \cosh (t-\lambda) \, d\lambda = o$

$$[\text{Ans.} \quad Y(t) = c(1-t^2/2)]$$

(64) Find the Laplace transform of

$$F(t) = o \qquad o < t < \frac{a}{3}$$

$$= c \qquad \frac{a}{3} < t < \frac{2a}{3}$$

$$= o \qquad \frac{2a}{3} < t < a$$

$$F(t+a) = F(t)$$

$$\left[\text{Ans. } \frac{c(e^{-\frac{a}{3}s} - e^{-\frac{2}{3}as})}{s(1-e^{-as})}\right]$$

(65) Evaluate $\displaystyle\int_0^\infty \frac{\cos 2t - \cos 4t}{t} \, dt$ [Ans. log 2]

(66) Evaluate $\displaystyle\int_{-\infty}^\infty e^{-t} \, U\,(t-3)dt$ [Ans. e^{-3}]

(67) Find $L\left\{ te^{-3t} J_0 \,(t\,\sqrt{2})\right\}$ $\left[\text{Ans. } \frac{s+3}{(s^2+6s+11)^{3/2}}\right]$

(68) Find using an inversion integral $L^{-1}\left\{ \frac{5s-2}{s^2(s+2)(s-1)} \right\}$

[Ans. $t - 2 + e^t + e^{-2t}$]

242

(69) Solve the difference equation

$$Y'(t) + Y(t-1) = t^2$$

given that $Y(t) = 0$ for $t \leq 0$.

$$\left[\text{Ans:} \quad Y(t) = \frac{2t^3}{3!} - \frac{2(t-1)^4}{4}U(t-1) + \frac{2(t-2)^5}{5!}U(t-2)-... \right]$$

(70) Find using an inversion integral $L^{-1} \left\{ \dfrac{1}{s^2(s^2+16)} \right\}$

$$\left[\text{Ans.} \quad \frac{1}{64} \left(4t - \sin 4t \right) \right]$$

3. Electrical **Applications** of the Laplace Transformation

Applications 1.

Find the current I at time t in a circuit consisting of a resistance R and inductance L in series with a condenser of capacity C when a constant E.M.F is applied at time t = 0, the initial values of charge and current being zero.

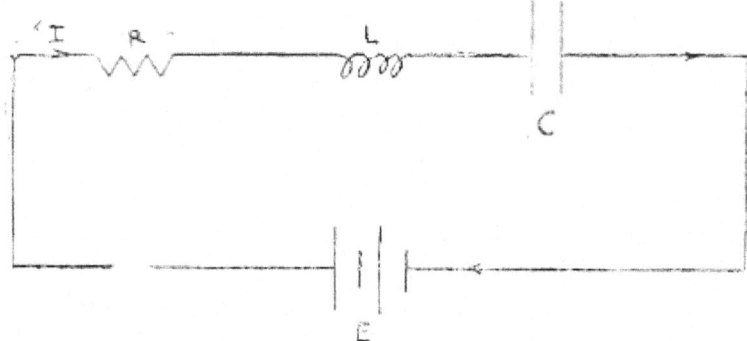

The differential equation satisfied by the current I is

$$IR + L\frac{dI}{dt} + \frac{1}{C}\int_0^t I(\tau)\, d\tau = E$$

with the initial conditions I(0) = 0, Q(0) = 0. The subsidiary equation is given by:

$$Ri(s) + Lsi(s) + \frac{i(s)}{Cs} = \frac{E}{s}$$

$$\therefore \left(R + Ls + \frac{1}{Cs} \right) i(s) = \frac{E}{s}$$

$$i(s) = \frac{E}{s\left(R + Ls + \frac{1}{Cs}\right)} = \frac{E}{Ls^2 + Rs + \frac{1}{C}} = \frac{E}{L} \cdot \frac{1}{s^2 + \frac{R}{L}s + \frac{1}{CL}}$$

$$= \frac{E}{L} \cdot \frac{1}{\left(s + \frac{R}{2L}\right)^2 + \frac{1}{CL} - \frac{R^2}{4L^2}} = \frac{E}{L} \cdot \frac{1}{\left(s + \frac{R}{2L}\right)^2 + n^2}$$

where $n^2 = \frac{1}{CL} - \frac{R^2}{4L^2}$

case (i) $n^2 + ve$ $I = \frac{E}{nL} e^{-\frac{R}{2L}t} \sin nt$

case (ii) $n^2 = 0$ $I = \dfrac{E}{nL} t e^{-\frac{R}{2L}t}$

case (iii) $n^2 -$ ve and is equal to $- \mu^2$

$$i(s) = \dfrac{E}{L} \cdot \dfrac{1}{\left(s + \dfrac{R}{2L}\right)^2 - \mu^2}$$

$\therefore$ $i = \dfrac{E}{\mu L} e^{-\frac{R}{2L}t} \sinh \mu t$

Application 2.

Find the current in a series circuit of resistance R, inductance L and capacity C in which an E.M.F. V is applied at t = 0, the initial values of the current I and charge Q on the condenser being respectively I_0 and Q_0.

The differential equation of the current is

$$L \dfrac{dI}{dt} + RI + \dfrac{Q}{C} = V$$

I and Q are connected by I = dQ/dt

The subsidiary equations are

$$L [s i(s) - I_0] + R i(s) + \dfrac{q(s)}{C} = v(s)$$

$$i(s) = sq(s) - Q_0$$

Eliminating q(s) we get

$$\left(Ls + R + \dfrac{1}{Cs}\right) i(s) = v(s) + LI_0 - \dfrac{Q_0}{Cs}$$

which on inversion gives the current I,

Application 3.

A battery of E M.F. E is connected at time t =0 to a series circuit of resistance R., inductance L and capacity C. The initial values of the current and charge are zero. If the battery is short-circuited at t = T find the current I at any instant t.

245

$$IR + L\frac{dI}{dt} + \frac{1}{C}\int_0^t I(\tau)\,d\tau = V$$

where $\quad V = E \qquad\qquad o < t < T,$

$$\qquad\qquad = o \qquad\qquad t > T$$

$$\therefore \quad v(s) = \frac{E}{s}\left(1 - e^{-Ts}\right)$$

$$= \frac{E}{Ls^2 + Rs + \dfrac{1}{C}}\left(1 - e^{-Ts}\right)$$

$$Ri(s) + Lsi(s) + \frac{1}{Cs}\,i(s) = \frac{E}{s}\left(1 - e^{-Ts}\right)$$

$$\therefore \quad i(s) = \frac{E}{s\left(R + Ls + \dfrac{1}{Cs}\right)}\left(1 - e^{-Ts}\right)$$

$$= \frac{E}{Ls^2 + Rs + \dfrac{1}{C}} - \frac{E}{Ls^2 + Rs + \dfrac{1}{C}}\,e^{-Ts}$$

$$= \frac{E}{L} \cdot \frac{1}{\left(s+\frac{R}{2L}\right)^2 + \frac{1}{CL} - \frac{R^2}{4L^2}} - \frac{E}{L} \cdot \frac{1}{\left(s+\frac{R}{2L}\right)^2 + \frac{1}{CL} - \frac{R^2}{4L^2}} e^{-Ts}$$

$$\therefore \ I = \frac{E}{nL} e^{-\frac{R}{2L}} \sin nt \quad o < t < T$$

$$= \frac{E}{nL} e^{-\frac{R}{2L}t} \sin nt - \frac{E}{nL} e^{-\frac{R}{2L}(t-T)} \sin n(t-T), \ t > T$$

$$\text{where} \ n^2 = \frac{1}{CL} - \frac{R^2}{4L^2} > o \ .$$

Application 4.

A periodic E. M. F. of resonance frequency E sin(nt) is applied at time t = 0 to a series circuit consisting of inductance L and capacity C where $n^2 = \frac{1}{CL}$.
Find the current the circuit at time t assuming initial zero current and charge.

$$L \frac{dI}{dt} + \frac{1}{C} \int_0^t I(\tau) d\tau = E \sin nt$$

with the conditions I(0) = 0, Q (0) = 0

$$Ls \ i(s) + \frac{1}{Cs} i(s) = \frac{En}{s^2+n^2}$$

$$\therefore \ i(s) = \frac{En}{\left(Ls + \frac{1}{Cs}\right)(s^2+n^2)} = \frac{Ens}{\left(Ls^2 + \frac{1}{C}\right)(s^2+n^2)}$$

$$= \frac{En}{L} \cdot \frac{s}{\left(s^2 + \frac{1}{CL}\right)(s^2 + n^2)} = \frac{En}{L} \frac{s}{(s^2 + n^2)^2}$$

and since $L^{-1} \left\{ \dfrac{s}{(s^2 + k^2)^2} \right\} = \dfrac{1}{2k} t \sin kt$ 1.13

then $i(s) = \dfrac{En}{L} \cdot \dfrac{1}{2n} t \sin nt$

$\qquad = \dfrac{E}{2L} t \sin nt$

Application 5.

Find the current at time t when a periodic E.M.F. E sin wt is applied at time $t = 0$ to an inductive resistance L, R the initial current being zero;

$$L \dfrac{dI}{dt} + RI = E \sin \omega t$$

with $I(0) = 0$.

$$(Ls + R) \, i(s) = \dfrac{E\omega}{s^2 + \omega^2}$$

$$\therefore \quad i(s) = \dfrac{E\omega}{(Ls + R)(s^2 + \omega^2)} = \dfrac{E\omega}{L} \cdot \dfrac{1}{\left(s + \dfrac{R}{L}\right)(s^2 + \omega^2)}$$

Put $\dfrac{R}{L} = \mu$

$$i(s) = \dfrac{E\omega}{L} \cdot \dfrac{1}{(s + \mu)(s^2 + \omega^2)}$$

$$= \dfrac{E\omega}{L(\mu^2 + \omega^2)} \left[\dfrac{1}{s + \mu} - \dfrac{s - \mu}{s^2 + \omega^2} \right]$$

$$\therefore \quad I(t) = \dfrac{E\omega}{L(\mu^2 + \omega^2)} \left[e^{-\mu t} - \cos \omega t + \dfrac{\mu}{\omega} \sin \omega t \right]$$

Application 6.

An E.M.F. E_1 for $o < t < T$ and E_2 for $t > T$ where E_1 and E_2 are constants is applied to a series circuit L, R, C. Find the current at any time t assuming zero initial current and charge.

$$L \dfrac{dI}{dt} + RI + \dfrac{1}{C} \int_o^t I(\tau) \, d\tau = V(t)$$

where $\quad V(t) = E_1 \qquad 0 < t < T$

$$= E_2 \qquad t > T$$

$$L\left\{V(t)\right\} = \frac{E_1}{s} - \frac{E_1 - E_2}{s} e^{-Ts}$$

$\therefore \quad \left(Ls + R + \dfrac{1}{Cs}\right) i(s) = \dfrac{E_1}{s} \dfrac{E_1 - E_2}{s} e^{-Ts}$

$$i(s) = \frac{E_1}{Ls^2 + Rs + \dfrac{1}{C}} - \frac{E_1 - E_2}{Ls^2 + Rs + \dfrac{1}{C}} e^{-Ts}$$

$$= \frac{E_1}{L} \cdot \frac{1}{s^2 + \dfrac{R}{L}s + \dfrac{1}{CL}} - \frac{E_1 - E_2}{L} \cdot \frac{1}{s^2 + \dfrac{R}{L}s + \dfrac{1}{CL}} e^{-Ts}$$

Now $\quad s^2 + \dfrac{R}{L}s + \dfrac{1}{CL} = \left(s + \dfrac{R}{2L}\right)^2 + \dfrac{1}{CL} - \dfrac{R^2}{4L^2}$

$= (s + \mu)^2 + n^2 \quad$ where $\mu = \dfrac{R}{2L}, \quad n^2 = \dfrac{1}{CL} - \dfrac{R^2}{4L^2} > 0$

$\therefore \quad i(s) = \dfrac{E_1}{L} \cdot \dfrac{1}{(s + \mu)^2 + n^2} - \dfrac{E_1 - E_2}{L} \cdot \dfrac{1}{(s + \mu)^2 + n^2} e^{-Ts}$

$\therefore \quad i(t) = \dfrac{E_1}{nL} e^{-\mu t} \sin nt \qquad 0 < t < T$,

$$= \frac{E_1}{nL} e^{-\mu t} \sin nt - \frac{E_1 - E_2}{nL} e^{-\mu(t-T)} \sin n(t-T), t > T.$$

Application 7.

The two circuits shown in the diagram are coupled by mutual inductance M.A constant E.M.F. E is applied to the primary at time t = 0 with zero initial conditions, the primary resistance being neglected. Find secondary current at any time t.

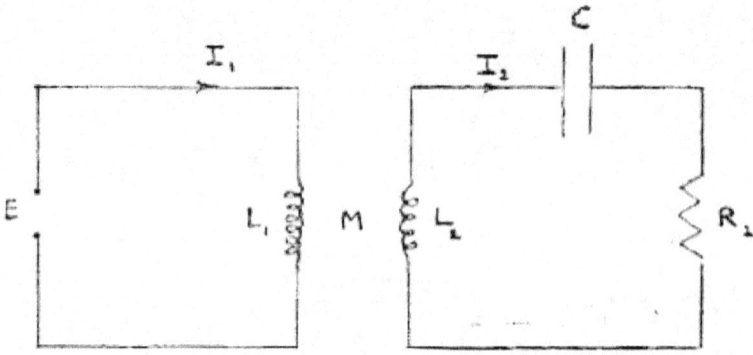

For the primary we have

$$L_1 \frac{dI_1}{dt} + M \frac{dI_2}{dt} = E \quad \ldots \quad (1)$$

with the initial condition $I1(0) = 0$.

For the secondary we have

$$I_2 R_2 + L_2 \frac{dI_2}{dt} + \frac{1}{C} \int_0^t I_2(\tau) \, d\tau + M \frac{dI_1}{dt} = 0 \quad \ldots \quad (2)$$

with the initial conditions $I2(0) = 0$, $Q(0) = 0$.
The subsidiary equations are

$$L_1 \, si_1 + Msi_2 = \frac{E}{s}$$

$$i_2 R_2 + L_2 si_1 + \frac{1}{Cs} \, i_2 + Msi_1 = 0$$

i.e $\quad L_1 si_1 + Msi_2 = \frac{E}{s} \quad \ldots \ldots (3)$

$$Msi_1 + (R_2 + L_2 s + \frac{1}{Cs}) \, i_2 = 0 \quad \ldots \quad (4)$$

Eliminating i_1 we get

$$i_2 = \frac{-ME}{(L_1L_2-M^2)s^2 + L_1R_2s + \dfrac{L_1}{C}}$$

$$= -\frac{ME}{L_1L_2-M^2} \cdot \frac{1}{s^2 + \dfrac{L_1 R_2 s}{L_1L_2-M^2} + \dfrac{L_1}{C(L_1L_2 - M^2)}}$$

$$= -\frac{ME}{L_1L_2-M^2} \cdot \frac{1}{(s+a)^2 + \beta^2}$$

Where

$$2a = \frac{L_1R_2}{L_1L_2-M^2}, \quad \beta^2 = \frac{L_1}{C(L_1L_2-M^2)} - a^2.$$

$$\therefore I_2 = -\frac{ME}{L_1L_2-M^2} e^{-at} \frac{1}{\beta} \sin \beta t$$

Application 8.

In the following network find the total current I at any instant t, the initial currents and charges are zero.

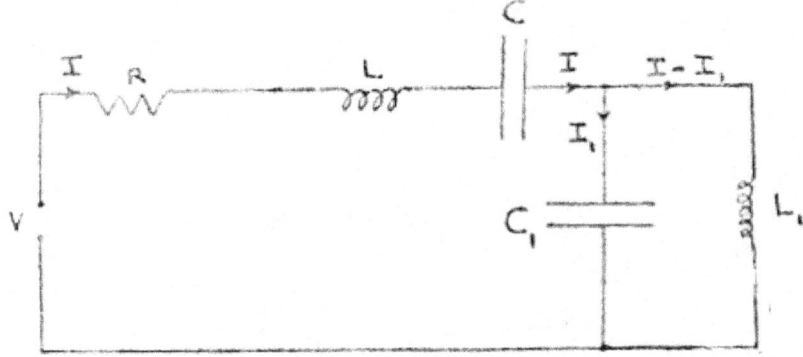

Applying Kirchhoff's voltage law to the circuit on the left which states that the impressed voltage equate the sum of the voltage drops across the elements in that circuit we get

$$IR + L\frac{dI}{dt} + \frac{1}{C} \int_0^t I(\tau) \, d\tau + \frac{1}{C_1} \int_0^t I_1(\tau) \, d\tau = V \quad \ldots \quad (i)$$

Again, applying the voltage law to the circuit on the right we get

251

$$L_1 \frac{d}{dt}(I-I_1) - \frac{1}{C_1} \int_0^t I_1(\tau)\, d\tau = 0 \quad \ldots \qquad (ii)$$

The subsidiary equations are :

$$Ri + Lsi + \frac{1}{Cs}i + \frac{1}{C_1 s} i_1 = v(s)$$

$$L_1(si - si_1) - \frac{1}{C_1 s} i_1 = 0$$

i.e $\quad \left(R + Ls + \frac{1}{Cs}\right) i + \frac{1}{C_1 s} i_1 = v(s) \quad \ldots \quad (iii)$

$$L_1 si - \left(L_1 s + \frac{1}{C_1 s}\right) i_1 = 0 \qquad \ldots \quad (iv)$$

Eliminating i_1 between (iii) and (iv) we get

$$i(s) = \frac{v(s) \cdot s(L_1 s^2 + C_1{}^{-1})}{p(s)}$$

where $p(s) = (L_1 s^2 + C_1{}^{-1})(Ls^2 + Rs + C^{-1} + C_1{}^{-1}) - C_1{}^{-2}$

Inverting we get the required current I.

<u>Application 9.</u>

In the following network, the switch S is closed at time t = o when both condensers are charged to voltage E. Find the current I.

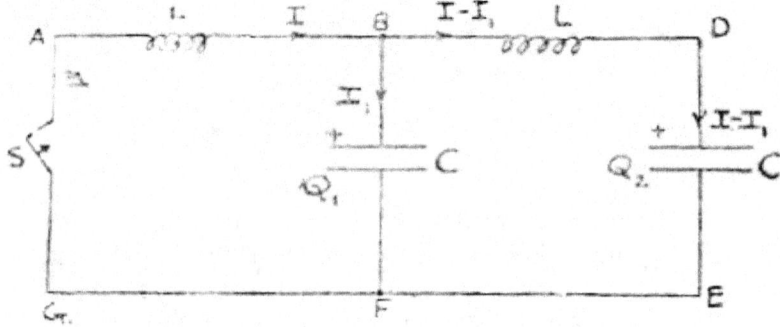

Applying the voltage law to the circuit ABFGA we get

$$L \frac{dI}{dt} + \frac{Q_1}{C} = 0 \quad \text{wher } I_1 = \frac{dQ_1}{dt}$$

Applying the voltage law to the circuit ABDEFGA we get

$$L \frac{dI}{dt} + L \frac{d}{dt} (I - I_1) + \frac{Q_2}{C} = o$$

where $\quad I - I_1 = \frac{dQ_2}{dt}$

$I(o) = I_1(o) = o, \quad Q_1(o) = Q_2(o) = EC$

$Lsi + \frac{q_1}{C} = o \quad$ where $i_1 = sq_1 - Q_1(o) = sq_1 - EC,$

$Lsi = L[si - si_1] + \frac{q}{C} = o$

where $i - i_1 = sq_2 - EC$

$\therefore \quad Lsi + \frac{1}{C} \left[\frac{i_1}{s} + \frac{EC}{s} \right] = o$

$Lsi + Lsi - Lsi_1 + \frac{1}{C} \left[\frac{i}{s} - \frac{i_1}{s} + \frac{EC}{s} \right] = o$

i.e $\quad Lsi + \frac{i_2}{Cs} = - \frac{E}{s} \qquad \dots \text{(i)}$

$$\left(2 Ls + \frac{1}{Cs} \right) i = \left(Ls + \frac{1}{Cs} \right) i_1 = - \frac{E}{s} \quad \dots \text{(ii)}$$

Eliminating i_1 between (i) and (ii) we get

$$i = - \frac{EC (CLs^2 + 2)}{C^2 L^2 s^4 + 3CLs^2 + 1}$$

Putting $n^2 = \frac{1}{CL}$ we get.

$$i = - \frac{E}{L} \frac{s^2 + 2n^2}{s^4 + 3n^2 s^2 + n^4}$$

$$= - \frac{E}{2L\sqrt{5}} \left[\frac{1 + \sqrt{5}}{s^2 + \frac{1}{2} (3 - \sqrt{5}) n^2} - \frac{1 - \sqrt{5}}{s^2 + \frac{1}{2} (3 + \sqrt{5}) n^2} \right]$$

$$\therefore I = - E \sqrt{\frac{C}{10L}} \left\{ \frac{1 + \sqrt{5}}{\sqrt{3 - \sqrt{5}}} \sin nt \left[\frac{1}{2} (3 - \sqrt{5}) \right]^{\frac{1}{2}} \right.$$

$$\left. - \frac{1 - \sqrt{5}}{\sqrt{3 + \sqrt{5}}} \sin nt \left[\frac{1}{2} (3 + \sqrt{5}) \right]^{\frac{1}{2}} \right\}$$

Application 10.

In the network shown below, determine the character of the current $I_1(t)$ assuming that each current is zero when the switch is closed.

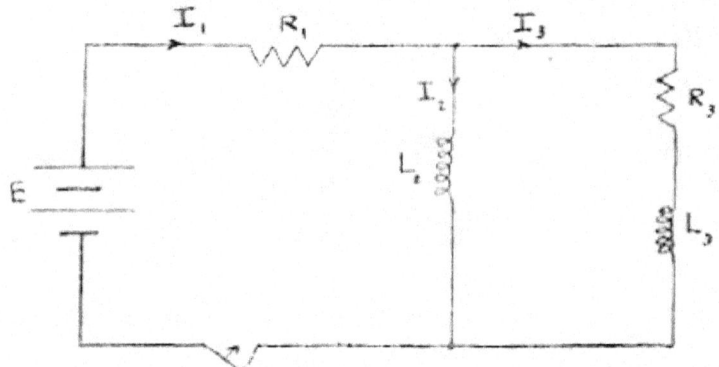

Since the algebraic sum of the currents at any junction is zero, then

$$I_1 - I_2 - I_3 = 0 \ . \ . \ . \ (1)$$

Applying the voltage law to the circuit on the left we get

$$R_1 I_1 + L_2 \frac{dI_2}{dt} = E \ . \ . \ . \ . \ (2)$$

Applying again the voltage low to the outside circuit we get

$$R_1 I_1 + R_3 I_3 + L_3 \frac{dI_3}{dt} = E \ . \ . \ . \ (3)$$

Transforming we get

$$i_1 - i_2 - i_3 = 0 \ . \ . \ . \ . \ (4)$$

$$R_1 i_1 + sL_2 i_2 = \frac{E}{s} \ . \ . \ . \ (5)$$

$$R_1 i_1 + (R_3 + sL_3) i_3 = \frac{E}{s} \ . \ (6)$$

$$\therefore \ i_1(s) = \frac{\begin{vmatrix} 0 & -1 & -1 \\ \dfrac{E}{s} & sL_2 & 0 \\ \dfrac{E}{s} & 0 & R_3 + sL_3 \end{vmatrix}}{\triangle} = \frac{E}{s} \ \frac{R_3 + s(L_2 + L_3)}{\triangle}$$

where

254

$$\Delta = \begin{vmatrix} 1 & -1 & -1 \\ R_1 & sL_2 & 0 \\ R_1 & 0 & R_3+sL_3 \end{vmatrix} = \begin{vmatrix} 1 & 0 & 0 \\ R_1 & sL_2+R_1 & R_1 \\ R_1 & R_1 & R_1+R_3+sL_3 \end{vmatrix}$$

i. o $\Delta = L_2 L_3 s^2 + (R_1 L_2 + R_3 L_2 + R_1 L_3)\,s + R_1 R_3$

Since we are interested in the factors of Δ, we consider the equation $\Delta = 0$. Since all coefficients of this equation are positive, hence it cannot have any positive roots. Its discriminant is

$$(R_1 L_2 + R_3 L_2 + R_1 L_3)^2 - 4L_2 L_3 R_1 R_3$$

which can be written

$$R_1^2 L_2^2 + 2R_1 L_2 (R_3 L_2 + R_1 L_3) + (R_3 L_2 - R_1 L_3)^2$$

which is positive. Hence the equation $\Delta = 0$ has two negative distinct roots $-a_1,\ -a_2$ (say),

$$\therefore\ \Delta = L_2 L_3 (s+a_1)\,(s+a_2)$$

$$\therefore\ i_1(s) = \frac{E}{s}\ \frac{R_3 + s\,(L_2+L_3)}{L_2 L_3\,(s+a_1)\,(s+a_2)}$$

$$= \frac{A_0}{s} + \frac{A_1}{s+a_1} + \frac{A_2}{s+a_2}$$

$$\therefore\ I_1(t) = A_0 + A_1 e^{-a_1 t} + A_2 e^{-a_2 t}$$

Application 11.

In the network shown below, derive the equations satisfied by the currents I1, I2, I3 and the chugs Q3 assuming that all initial currents and charges are zero and obtain the transformed equations.

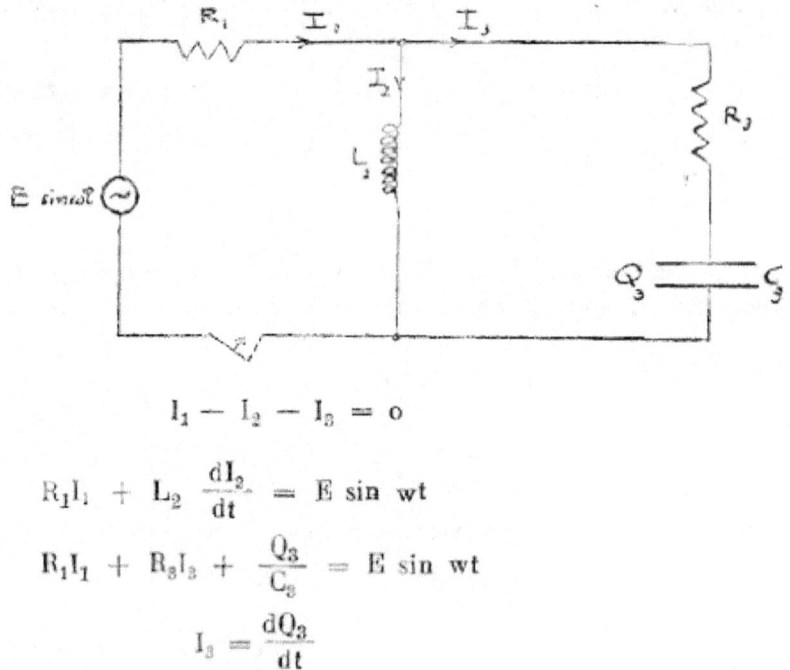

$$I_1 - I_2 - I_3 = 0$$

$$R_1 I_1 + L_2 \frac{dI_2}{dt} = E \sin wt$$

$$R_1 I_1 + R_3 I_3 + \frac{Q_3}{C_3} = E \sin wt$$

$$I_3 = \frac{dQ_3}{dt}$$

Initial conditions are

$$I_1(0) = I_2(0) = i_3(0) = 0, \quad Q_3(0) = 0$$

Transforming we get

$$i_1 - i_2 - i_3 = 0$$

$$R_1 i_1 + s L_2 i_2 = \frac{E\omega}{s^2 + \omega^2}$$

$$R_1 i_1 + R_3 i_3 + \frac{q_3}{C_3} = \frac{E\omega}{s^2 + \omega^2}$$

$$i_3 = s q_3 \quad \text{or} \quad q_3 = \frac{i_3}{s}$$

The solution can then be completed as in the previous example,

Application 12.

An impulsive E.M.F. $E_0 \delta(t)$ is applied at t = 0 to an L, C, R, circuit in series with zero initial currents and charges. Find the current at *any* instant t,

256

$$L\frac{dI}{dt} + IR + \frac{1}{C}\int_0^t I(\tau)\, d\tau = E_0\delta(t)$$

$$\therefore \left(Ls + R + \frac{1}{Cs}\right) i(s) = E_0$$

$$i(s) = \frac{E_0 s}{Ls^2 + Rs + \frac{1}{C}}$$

$$= \frac{E_0}{L} \cdot \frac{s}{s^2 + \frac{R}{L}s + \frac{1}{CL}}$$

$$= \frac{E_0}{L} \cdot \frac{s}{\left(s + \frac{R}{2L}\right)^2 + \frac{1}{CL} - \frac{R^2}{4L^2}}$$

$$= \frac{E_0}{L} \frac{s + \frac{R}{2L}}{\left(s + \frac{R}{2L}\right)^2 + \frac{1}{CL} - \frac{R^2}{4L^2}}$$

$$- \frac{E_0}{L} \frac{\frac{R}{2L}}{\left(s + \frac{R}{2L}\right)^2 + \frac{1}{CL} - \frac{R^2}{4L^2}}$$

$$\therefore \quad I(t) = \frac{E_0}{L} e^{-\frac{R}{2L}t} \cos nt - \frac{E_0 R}{2L^2 n} e^{-\frac{R}{2L}t} \sin nt$$

where

$$n^2 = \frac{1}{CL} - \frac{R^2}{4L^2} > 0.$$

4. Dynamical Applications of Laplace Transforms

Application 1

A particle is projected vertically upwards at time t = 0 with velocity v_0 from the origin under the action of gravity and a resistance equal to 2 km times the velocity. Find its displacement at any instant t.

The equation of motion of the mass is

$$m\ddot{x} = -2km\dot{x} - mg$$

with the initial conditions

$$x(o) = o, \quad \dot{x}(o) = v_o$$

$$\therefore \quad \ddot{x} + 2k\dot{x} = -g$$

$$s^2\bar{x} - sx(o) - \dot{x}(o) + 2k[s\bar{x} - x(o)] = -\frac{g}{s}$$

$$(s^2 + 2ks)\bar{x} = v_o - \frac{g}{s}$$

$$\therefore \quad \bar{x} = \frac{v_o}{s(s+2k)} - \frac{g}{s^2(s+2k)}$$

$$= \frac{v_o}{2k}\left[\frac{1}{s} - \frac{1}{s+2k}\right] - g\left[\frac{1}{2ks^2} - \frac{1}{4k^2s} + \frac{1}{4k^2(s+2k)}\right]$$

$$\therefore \quad x = \frac{v_o}{2k}\left(1 - e^{-2kt}\right) - g\left[\frac{t}{2k} - \frac{1}{4k^2} + \frac{1}{4k^2}e^{-2kt}\right]$$

$$\text{i.e} \quad x(t) = -\frac{gt}{2k} + \frac{g + 2kv_o}{4k^2}\left(1 - e^{-2kt}\right)$$

Application 2.

A particle of mass m moves in a vertical plane under the action of gravity and a force directed towards the origin and equal to μr where r is the distance of the particle from the origin. If the particle is projected from the point (a, 0) vertically upwards with velocity v_o, find the coordinates of the particle at any instant t.

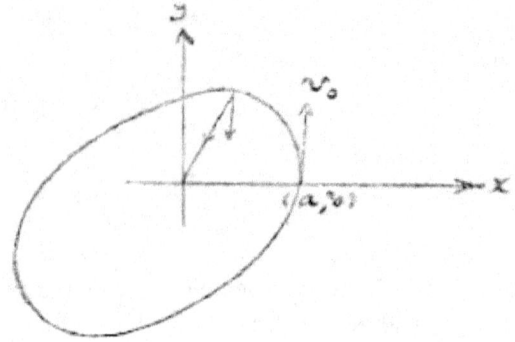

$$m\ddot{x} = -\mu\dot{x}, \quad m\ddot{y} = -\mu\dot{y} - mg$$

Put $\dfrac{\mu}{m} = n^2$

$$\therefore \quad \ddot{x} + n^2x = 0, \quad \ddot{y} + n^2y = -g$$

The initial conditions are

$$x(o) = a, \quad y(o) = 0, \quad \dot{x}(o) = 0, \quad \dot{y}(o) = v_o$$

$$s^2\bar{x}(s) - sx(o) - \dot{x}(o) + n^2\bar{x}(s) = 0$$

$$(s^2 + n^2)\bar{x} = as$$

$$\therefore \quad \bar{x} = \frac{as}{s^2 + n^2} \qquad \therefore \quad x(t) = a\cos nt$$

$$s^2\bar{y}(s) - sy(o) - \dot{y}(o) + n^2\bar{y}(s) = -\frac{g}{s}$$

$$s^2\bar{y} - v_o + n^2\bar{y} = -\frac{g}{s}$$

$$(s^2 + n^2)\bar{y} = v_o - \frac{g}{s}$$

$$\bar{y} = \frac{v_o}{s^2 + n^2} - \frac{g}{s(s^2 + n^2)}$$

Now $L^{-1}\left\{\dfrac{1}{s(s^2 + n^2)}\right\} = \dfrac{1}{n^2}(1 - \cos nt)$

$$\therefore \quad y(t) = \frac{v_o}{n}\sin nt - \frac{g}{n^2}(1 - \cos nt)$$

Application 3.

A particle of mass m moves in a straight line under a resistance $2\mu m$ times the velocity and a restoring force $m\lambda$ times the displacement where $\lambda > \mu^2$. If it is projected at time t = 0o with velocity v_o at distance x_o; from its equilibrium position, find its displacement at any subsequent instant t.

Equation of motion of the mass is

$$m\ddot{x} = -2pm\dot{x} - m\lambda x$$

i.e $\quad \ddot{x} + 2p\dot{x} + \lambda x = 0$

with the initial conditions

$$x(0) = x_0 \quad , \quad \dot{x}(0) = v_0$$

$$[s^2\bar{x} - sx(0) - \dot{x}(0)] + 2\mu[s\bar{x} - x(0)]\lambda\bar{x} = 0$$

$$\therefore \quad (s^2 + 2\mu s + \lambda)\bar{x} = sx_0 + v_0 + 2\mu x_0$$

$$\bar{x} = \frac{s\lambda_0 + v_0 + 2\mu x_0}{(s+\mu)^2 + \lambda - \mu^2}$$

Put $\lambda - \mu^2 = n^2$

$$\therefore \quad \bar{x} = \frac{x_0(s+\mu) + v_0 + \mu x_0}{(s+\mu)^2 + n^2}$$

$$= \frac{x_0(s+\mu)}{(s+\mu)^2 + n^2} + \frac{v_0 + \mu x_0}{(s+\mu)^2 + n^2}$$

$$\therefore \quad x(t) = x_0 e^{-\mu t}\cos nt + \frac{v_0 + \mu x_0}{n}e^{-\mu t}\sin nt$$

$$= \frac{1}{n}e^{-\mu t}[nx_0 \cos nt + (x_0 + \mu x_0)\sin nt]$$

Application 4

A spring of stiffness k is placed on a smooth horizontal table. One end of the spring is fixed at a point O on the table and to the other end is attached a mass m. The system is initially at rest with the spring unstretched .A constant force F is applied to the mass for a time t_0 and then removed. Find the mass at any time t.

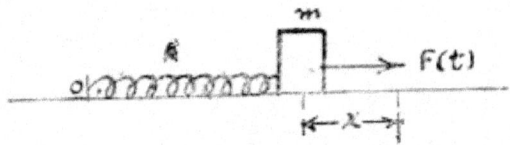

Equation of motion of the mass is

$$m\ddot{x} = -kx + F(t)$$

where $F(t) = F_0 \qquad 0 < t < t_0$

$$= 0 \qquad t > t_0$$

The initial conditions are

$$x(0) = \dot{x}(0) = 0$$

$$\ddot{x} = -\frac{k}{m}x + \frac{1}{m}F(t).$$

Put $\dfrac{k}{m} = n^2$

$$\therefore \ddot{x} + n^2 x = \frac{1}{m}F(t)$$

Transforming and noticing that

$$L\left\{F(t)\right\} = \frac{F_o}{s}(1-e^{-t_o s})$$

We get

$$(s^2 + n^2)\,x = \frac{F_o}{ms}(1-e^{-t_o s})$$

$$\bar{x} = \frac{F_o}{ms(s^2+n^2)}(1-e^{-t_o s})$$

Now $L^{-1}\left\{\dfrac{1}{s(s^2+n^2)}\right\} = \dfrac{1}{n^2}(1-\cos nt) = \dfrac{2}{n^2}\sin^2\dfrac{nt}{2}$

$$\therefore \quad x(t) = \frac{2F_o}{mn^2}\sin^2\frac{nt}{2} \qquad\qquad 0 < t < t_o$$

$$= \frac{2F_o}{mn^2}\sin^2\frac{nt}{2} - \frac{2F_o}{mn^2}\sin^2\frac{n}{2}(t-t_o) \qquad t > t_o$$

$$\text{i.e} \quad x(t) = \frac{2F_o}{k}\sin^2\frac{nt}{2} \qquad\qquad 0 < t < t_o$$

$$= \frac{2F_o}{k}\left[\sin^2\frac{nt}{2} - \sin^2\frac{n}{2}(t-t_o)\right] \qquad t > t_o$$

Application 5

Two particles each of mass m are connected by a spring of stiffness k and they are free to move in a straight line on a smooth horizontal table. At time t = o when both particles are at rest and the spring is unstrained a constant force P is applied to one of them in the direction towards the other particle. Find the displacement of the other particle from its initial position at any subsequent instant t.

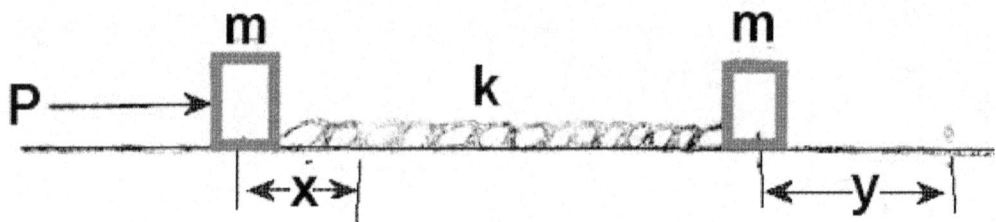

The equations of motion of the two masses are

$$m\ddot{x} = k(y-x) + P$$

$$m\ddot{y} = -k(y-x)$$

Putting $k/m = n^2$ we get

$$\ddot{x} = n^2(y-x) + \frac{P}{m}$$

$$\ddot{y} = -n^2(y-x)$$

The initial conditions are

$$x(o) = y(o) = \dot{x}(o) = \dot{y}(o) = o$$

$$\therefore \quad (s^2 + n^2)\bar{x} - n^2\bar{y} = \frac{P}{ms}$$

$$-n^2\bar{x} + (s^2 + n^2)\bar{y} = o$$

Eliminate $\bar{x}$

$$\therefore \quad [(s^2+n^2)^2 - n^4]\bar{y} = \frac{n^2P}{ms}$$

$$\therefore \quad s^2(s^2+2n^2)\bar{y} = \frac{n^2P}{ms}$$

$$\bar{y} = \frac{n^2P}{m} \cdot \frac{1}{s^3(s^2+2n^2)}$$

$$= \frac{P}{4m}\left[\frac{2}{s^3} - \frac{1}{n^2s} + \frac{s}{n^2(s^2+2n^2)}\right]$$

$$\therefore \quad y(t) = \frac{P}{4m}\left[t^2 - \frac{m}{k} + \frac{m}{k}\cos t\sqrt{\frac{2k}{m}}\right]$$

Application 6

262

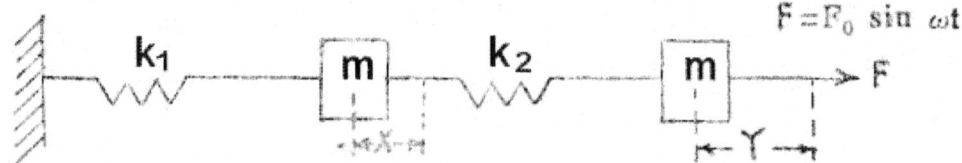

Two masses are equal each being equal to m. The stiffnesses k_1, k_2 of the two springs are connected by the relation $k_1 / k_2 = 3 / 2$. The system is initially at rest with the two springs unstrained. A periodic force $F = F_0 \sin \omega t$ where $\omega^2 = k_2/m$ is applied to the right mass. Find the displacement $X(t)$ of the left mass at any instant t.

Solution:

$$m\ddot{X} = k_2(Y-X) - k_1 X$$

$$m\ddot{Y} = -k_2(Y-X) + F_0 \sin \omega t$$

$$\therefore \ddot{X} = \omega^2(Y-X) - \frac{3}{2}\omega^2 X$$

$$\ddot{Y} = -\omega^2(Y-X) + \frac{F_0}{m}\sin \omega t$$

Initial conditions are

$$X(o) = Y(o) = \dot{X}(o) = \dot{Y}(o) = o$$

$$\ddot{X} + \frac{5}{2}\omega^2 X - \omega^2 Y = o$$

$$-\omega^2 X + \ddot{Y} + \omega^2 Y = \frac{F_0}{m}\sin \omega t$$

$$\therefore (D^2 + \frac{5}{2}\omega^2) X - \omega^2 Y = o$$

$$-\omega^2 X + (D^2 + \omega^2) Y = \frac{F_0}{m}\sin \omega t$$

$$\therefore (s^2 + \frac{5}{2}\omega^2) x - \omega^2 y = o$$

$$-\omega^2 x + (s^2 + \omega^2) y = \frac{F_0 \omega}{m} \cdot \frac{1}{s^2 + \omega^2}$$

Eliminating y we get

$$[(s^2 + \frac{5}{2}\omega^2)(s^2 + \omega^2) - \omega^4] x = \frac{F_0 \omega^3}{m} \cdot \frac{1}{s^2 + \omega^2}$$

$$[s^4 + \frac{7}{2}s^2\omega^2 + \frac{3}{2}\omega^4] x = \frac{F_0 \omega^5}{k_2} \cdot \frac{1}{s^2 + \omega^2}$$

$$\therefore \quad (s^2 + \tfrac{1}{2}\,\omega^2)\,(s^2 + 3\omega^2)\,x = \frac{F_0\omega^5}{k_2} \cdot \frac{1}{s^2+\omega^2}$$

$$x(s) = \frac{F_0\omega^5}{k_2} \cdot \frac{1}{(s^2+\tfrac{1}{2}\omega^2)(s^2+3\omega^2)(s^2+\omega^2)}$$

$$= \frac{F_0\omega^5}{k_2}\left[\frac{4}{5\omega^4} \cdot \frac{1}{s^2+\tfrac{1}{2}\omega^2} + \frac{1}{5\omega^4(s^2+3\omega^2)} - \frac{1}{\omega^4(s^2+\omega^2)}\right]$$

$$= \frac{F_0\omega}{5k_2}\left[\frac{4}{s^2+\tfrac{1}{2}\omega^2} + \frac{1}{s^2+3\omega^2} - \frac{5}{s^2+\omega^2}\right]$$

$$\therefore X(t) = \frac{F_0\omega}{5k_2}\left[\frac{4\sqrt{2}}{\omega}\sin\sqrt{\tfrac{1}{2}}\,\omega t + \frac{1}{\sqrt{3}\,\omega}\sin\sqrt{3}\,\omega t - \frac{5}{\omega}\sin\omega t\right]$$

i.e $15\,k_2 X(t) = F_0[12\sqrt{2}\,\sin(\sqrt{\tfrac{1}{2}}\,\omega t) + \sqrt{3}\,\sin(\sqrt{3}\,\omega t) - 15\sin\omega t]$

Application 7

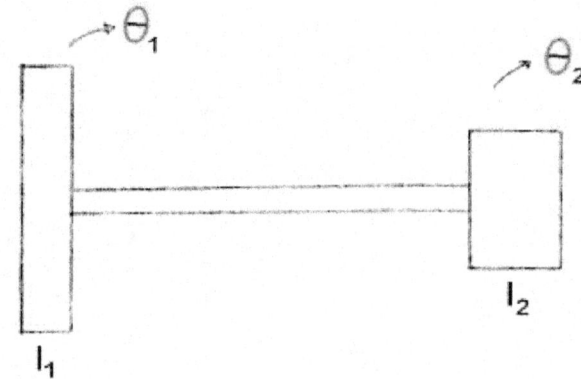

Two flywheels of moments of inertia I1 and I2 are connected by an elastic shaft of torsional stiffness A i.e., the couple per radian relative twist of the flywheels is A. The whole system is rotating with a constant angular velocity ω when at time t = 0, a constant retarding couple P is applied to the wheel I1 , Find the angular velocity of the wheel I2 at any instant t.

Solution:

     Let θ_1 be the angular displacement of the flywheel I_1 and θ_2 that of I_2 .

Equations of motion are

$$I_1\ddot{\theta_1} = \lambda(\theta_2 - \theta_1) - P$$

$$I_2\ddot{\theta_2} = -\lambda(\theta_2 - \theta_1)$$

Initial conditions can be taken as

264

$$\theta_1(0) = \theta_2(0) = 0 \quad , \quad \dot{\theta}_1(0) = \dot{\theta}_2(0) = \omega$$

$$\therefore \quad I_1 [s^2\bar{\theta}_1 - s\theta_1(0) - \dot{\theta}_1(0)] = \lambda(\bar{\theta}_2 - \bar{\theta}_1) - \frac{P}{s}$$

$$I_2 [s^2\bar{\theta}_2 - s\theta_2(0) - \dot{\theta}_2(0)] = -\lambda(\bar{\theta}_2 - \bar{\theta}_1)$$

$$(I_1 s^2 + \lambda)\,\bar{\theta}_1 - \lambda\bar{\theta}_2 = I_1\omega - \frac{P}{s}$$

$$-\lambda\bar{\theta}_1 + (I_2 s^2 + \lambda)\bar{\theta}_2 = I_2\omega$$

Eliminate $\bar{\theta}_1$

$$[(I_1 s^2 + \lambda)(I_2 s^2 + \lambda) - \lambda^2]\bar{\theta}_2 = \lambda\left(I_1\omega - \frac{P}{s}\right) + I_2\omega(I_1 s^2 + \lambda)$$

$$\therefore \quad s^2[I_1 I_2 s^2 + \lambda(I_1 + I_2)]\bar{\theta}_2 = \omega[I_1 I_2 s^2 + \lambda(I_1 + I_2)] - \frac{\lambda P}{s}$$

$$\bar{\theta}_2 = \frac{\omega}{s^2} - \frac{\lambda P}{s^2[I_1 I_2 s^2 + \lambda(I_1 + I_2)]}$$

Let $\phi = \dot{\theta}_2$

$$\bar{\phi} = s\bar{\theta}_2 - \theta_2(0) = \frac{\omega}{s} - \frac{\lambda P}{s^2[I_1 I_2 s^2 + \lambda(I_1 + I_2)]}$$

$$\frac{\omega}{s} - \frac{\lambda P}{I_1 I_2 s^2(s^2 + n^2)}$$

where $n^2 = \dfrac{\lambda(I_1 + I_2)}{I_1 I_2}$

$$\therefore \quad \bar{\phi} = \frac{\omega}{s} - \frac{\lambda P}{I_1 I_2 n^2}\left[\frac{1}{s^2} - \frac{1}{s^2 + n^2}\right]$$

$$= \frac{\omega}{s} - \frac{P}{I_1 + I_2}\left[\frac{1}{s^2} - \frac{1}{s^2 + n^2}\right]$$

$$\therefore \quad \phi = \omega - \frac{Pt}{I_1 + I_2} + \frac{P}{n(I_1 + I_2)}\sin nt$$

Application 8

In previous application, let the retarding couple P be applied for time T only.

Solution:

Here the transform of P is

$$\frac{P}{s}\left(1-e^{-Ts}\right)$$

instead of P/s in the last problem.

$$\text{Thus } \phi = \frac{\omega}{s} - \frac{P(1-e^{-Ts})}{I_1+I_2}\left[\frac{1}{s^2}-\frac{1}{s^2+n^2}\right]$$

$$\therefore \ \phi = \omega - \frac{Pt}{I_1+I_2} + \frac{P}{n(I_1+I_2)}\sin nt \qquad 0<t<<T$$

$$= \omega - \frac{Pt}{I_1+I_2} + \frac{P}{n(I_1+I_2)}\sin nt + \frac{P(t-T)}{I_1+I_2}$$

$$- \frac{P}{n(I_1+I_2)}\sin n(t-T) , \ t > T.$$

Application 9

The tautochrone is the shape taken by a smooth wire through the origin in the vertical x-y plane, such that when a bead is constrained to slide on it, time of descent of the bead from any point on the wire at which it is at rest to the origin which is its lowest point is constant. i.e., independent of the starting point.

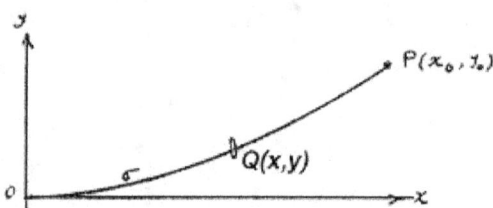

Solution:

Let $P(x0,yo)$ be the starting print i.e., the bead is at rest at P and starts sliding and let $Q(x,y)$ be any point on the wire between O and P where O is the origin taken at the lowest point of the wire. Let $\sigma =$ arc OQ measured from O and m = mass of bead. The loss of potential energy when the bead falls through the vertical distance yo - y is equal to the gain of the kinetic energy.

$$\therefore \ \tfrac{1}{2} m \left(\frac{d\sigma}{dt}\right)^2 = mg \ (y_0-y)$$

$$\therefore \ \frac{d\sigma}{dt} = -\sqrt{2g(y_0-y)}$$

The negative sign is taken since σ decreases as t increases. Let T be the time of descent from P to O, then

$$T = - \int_{y_o}^{o} \frac{d\sigma}{\sqrt{2g(y_0-y)}}$$

$$\text{i.e } T = \int_{o}^{y_0} \frac{d\sigma}{\sqrt{2g(y_0-y)}}$$

Now when the equation of the curve is known, we can express the length of the arc σ in terms of y and consequently the differential of the arc $d\sigma$ can be expressed in terms of y and dy — Hence we can put $d\sigma = F(y) dy$

$$\therefore T\sqrt{2g} = \int_{o}^{y_0} \frac{F(y)dy}{\sqrt{(y_0-y)}}$$

This is an integral equation of the convolution type in which the unknown function F(y) has to be determined such that T is constant that is independent of yo. Now the integral equation can be written

$$T\sqrt{2g} = F(y) * y^{-\frac{1}{2}}$$

Transforming, we get

$$\frac{T\sqrt{2g}}{s} = f(s)\sqrt{\frac{\pi}{s}}$$

$$\therefore f(s) = T\sqrt{\frac{2g}{\pi}} \cdot \frac{1}{\sqrt{s}}$$

Inverting we get

$$F(y) = T\sqrt{\frac{2g}{\pi}} \cdot \frac{1}{\sqrt{\pi y}}$$

$$= \frac{T}{\pi}\sqrt{2g} \cdot \frac{1}{\sqrt{y}}$$

Now $\frac{d\sigma}{dy} = \frac{\sqrt{dx^2+dy^2}}{dy} = \sqrt{1 + \left(\frac{dx}{dy}\right)^2} = F(y)$

$$= \frac{T}{\pi}\sqrt{2g} \cdot \frac{1}{\sqrt{y}}$$

i.e $\sqrt{1 + \left(\frac{dx}{dy}\right)^2} = \frac{T}{\pi}\sqrt{2g} \cdot \frac{1}{\sqrt{y}}$

$$1 + \left(\frac{dx}{dy}\right)^2 = \frac{2gT^2}{\pi^2} \cdot \frac{1}{y}$$

Put $c = \dfrac{2gT^2}{\pi^2}$

$$\therefore \quad 1 + \left(\frac{dx}{dy}\right)^2 = \frac{c}{y}$$

i.e $\left(\dfrac{dx}{dy}\right)^2 = \dfrac{c-y}{y}$ $\therefore \quad \dfrac{dx}{dy} = \sqrt{\dfrac{c-y}{y}}$

i.e $x = \displaystyle\int \sqrt{\dfrac{c-y}{y}} \; dy + d$

Put $y = c \sin^2 \theta$

$$x = \int \sqrt{\frac{c \cos^2\theta}{c \sin^2\theta}} \cdot 2c \sin\theta \cos\theta \, d\theta + d$$

$$= 2c \int \cos^2\theta \, d\theta + d$$

$$= c \int (1 + \cos 2\theta) \, d\theta + d$$

$$= c \left[\theta + \frac{\sin 2\theta}{2} \right] + d$$

$$= \frac{c}{2} [2\theta + \sin 2\theta] + d$$

and $y = \dfrac{c}{2} [1 - \cos 2\theta]$

Now since the curve passes through origin $d = 0$ and the parametric equations of the curve can be put in the form

$$x = a [\phi + \sin \phi]$$
$$y = a [1 - \cos \phi]$$

where $a = \dfrac{c}{2} = \dfrac{gT^2}{\pi^2}$ and $\phi = 2\theta$

These are the parametric equations of a cycloid. Hence the wire assumes the shape of a

cycloid in which the radius of the generating circle is $\dfrac{gT^2}{\pi^2}$.

Application 10

268

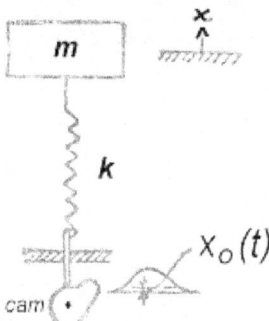

The figure shows a spring - mass system, the lower end of which undergoes a motion $x_o(t)$ prescribed by a cam.

The equation of motion of the mass m is

$$m\ddot{x} = k \left[x_o(t) - x \right]$$

$$\therefore \quad \ddot{x} = \frac{k}{m} \left[x_o(t) - x \right]$$

Put $w^2 = k/m$

$$\therefore \quad \ddot{x} + w^2 x = w^2 x_o(t)$$

If the system is started from rest, the subsidiary equation for the displacement of m is

$$(s^2 + w^2) \, x(s) = w^2 x_o(s)$$

i.e $\quad \bar{x}(s) = \dfrac{w^2 \bar{x_o}(s)}{s^2 + w^2}$

When $x_o(t)$ is specified. $\bar{x_o}(s)$ can be determined and consequently $\bar{x}(s)$ and x(t).

To obtain the force exerted on in by the spring in which case we are concerned with the relative motion $y = x_o(t) - x$ of the ends of the spring, we have

$$F = ky = m\ddot{x}$$

The subsidiary equation of which is

$$\bar{F}(s) = ms^2 \, \bar{x}(s) = \frac{m\omega^2 s^2 \bar{x_o}(s)}{s^2 + w^2}$$

Application 11

A uniform rod of length 2a and mass m is at rest on a smooth horizontal table, At time t = 0, it is set in motion by a blow of impulse P at one end of the rod perpendicular to the rod. Find and solve its equation of motion.

Solution:

Let x be the linear displacement of the middle point of the rod and θ the angular displacement of the rod about its centre.

$$m \ddot{x} = P \, \delta(t)$$

$$\tfrac{1}{3} ma^2 \ddot{\theta} = P a \, \delta(t)$$

Transforming we get

$$ms^2 \bar{x} = P, \quad \tfrac{1}{3} mas^2 \bar{\theta} = P$$

i.e

$$\bar{x} = \frac{P}{ms^2}, \quad \bar{\theta} = \frac{3P}{mas^2}$$

$$\therefore \quad x(t) = \frac{P}{m} t, \quad \theta(t) = \frac{3 Pt}{ma}$$

5. STRUCTURAL APPLICATIONS

5.1. Deflection of beams

The differential equation for the deflection y of a team is

$$EI \frac{d^2y}{dx^2} = -M$$

where M is the bending moment at a point x of the beam, F is its Young's modulus and I the moment of inertia of the cross section of the beam about its neutral axis.

For transverse loading w per unit length of the beam, the above differential equation becomes:

$$EI \frac{d^4y}{dx^4} = w$$

provided E and I are constants.

A concentrated load W at x = a may be considered as a distributed load w per unit length of the beam such that

$$w = W\delta(x-a) = WU'(x-a)$$

where δ is the Dirac delta function. Its transform is

$$\bar{w} = We^{-as}$$

A constant load wo per unit length in o < x < a and zero load for x > a may be written

$$w = w_o \left[1 - U(x-a) \right]$$

and

$$\bar{w} = \frac{w_o}{s} \left(1 - e^{-as} \right)$$

The transform of a couple

$$\mu \, \delta' \, (x-a) \quad \mu \, U'' \, (x-a)$$

of moment μ applied at $x = a$ is $\mu s e^{-as}$

Now consider the general beam equation

$$EI \, \frac{d^4 y}{dx^4} = f(x)$$

or

$$\frac{d^4 y}{dx^4} = \frac{1}{EI} \, f(x)$$

where $f(x)$ represents the load per unit length at a point x of the beam.

The subsidiary equation is given by

$$s^4 \, \bar{y}(s) - s^3 y(o) - s^2 y'(o) - s y''(o) - y'''(o) = \frac{1}{EI} \, \bar{f}(s)$$

i.e $\bar{y}(s) = \dfrac{y(o)}{s} + \dfrac{y'(o)}{s^2} + \dfrac{y''(o)}{s^3} + \dfrac{y'''(o)}{s^4} + \dfrac{1}{EI} \, \dfrac{\bar{f}(s)}{s^4}$

Inverting we get

$$y(x) = y(o) + y'(o) \, x + y''(o) \frac{x^2}{2!} + y'''(o) \frac{x^3}{3!} + \frac{1}{EI} \, L^{-1} \left\{ \frac{\bar{f}(s)}{s^4} \right\}$$

In practical problems some of the quantities $y(0)$, $y'(0)$, $y''(0)$. $y'''(0)$ are known if we are given at $x = 0$ the deflection or slope or bending moment or shearing force. The remaining quantities are determined from conditions at other points of the beam.

The following are applications of the above principles.

Application 1

A beam is hinged at its ends $x = o$ and $x = l$. It carries a uniformly distributed lead w perr unit length. Find the static deflection at any point on the beam.

$$EI \, \frac{d^4 y}{dx^4} = w_o \qquad \therefore \quad \frac{d^4 y}{dx^4} = \frac{w_o}{EI}$$

$$s^4 \, \bar{y} - s^3 y(o) - s^2 y'(o) - s y''(o) - y'''(o) = \frac{w_o}{EIs}$$

$$y(o) = y''(o) = o \text{ and put } y'(o) = A \text{ and } y'''(o) = B$$

$$s^4 \, \bar{y} = As^2 + B + \frac{w_o}{EIs}$$

$$\therefore \quad \bar{y} = \frac{A}{s^2} + \frac{B}{s^4} + \frac{w_o}{EIs^5}$$

$$y = Ax + B\frac{x^3}{3!} + \frac{w_o}{EI}\frac{x^4}{4!}$$

we have now to determine A and B from the other two conditions namely $y(l) = y''(l) = 0$,

$$0 = Al + B\frac{l^3}{6} + \frac{w_o}{EI}\frac{l^2}{24}$$

$$0 = Bl + \frac{w_o}{EI}\frac{l^2}{2}$$

$$\therefore \quad B = -\frac{w_o}{EI}\frac{l}{2} \quad , \quad A = \frac{w_o l^3}{24EI}$$

$$\therefore \quad y = \frac{w_o l^3}{24EI}x - \frac{w_o l}{12EI}x^3 + \frac{w_o}{24EI}x^4$$

$$= \frac{w_o}{24EI}(l^3x - 2lx^3 + x^4)$$

$$= \frac{w_o}{24EI}x(l - x)(l^2 + lx - x^2)$$

Application 2

A beam of length 2 l has both its ends built in. It carries a load w(x) per wilt length such that

$$W(x) = w_o \qquad 0 < x < l$$

$$= 0 \qquad l < x < 2l$$

Find the static deflection y at any point x.

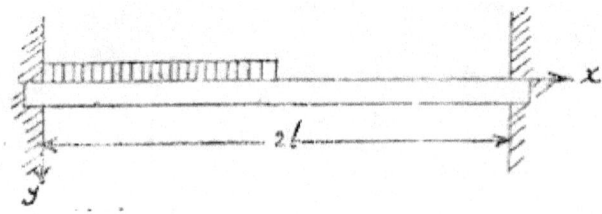

$$\frac{d^4y}{dx^4} = a\ w(x) \quad \text{where} \quad a = \frac{1}{EI}$$

$$s^4\bar{y}(s) - s^3y(o) - s^2y'(o) - s\ y''(o) - y'''(o)$$

$$= \frac{aw_o}{s}\ (1-e^{-ls})$$

Now y(0) = y'(0) = 0 and let y"(0) = A, y"'(0) = B

$$\therefore\quad s^4\bar{y} = As + B + \frac{aw_o}{s}\ (1-e^{-ls})$$

$$\bar{y} = \frac{A}{s^3} + \frac{B}{s^4} + \frac{aw_o}{s^5}\ (1-e^{-ls})$$

$$y = \frac{Ax^2}{2!} + \frac{Bx^3}{3!} + aw_o\ \frac{x^4}{4!} \qquad\qquad o < x < l,$$

$$= \frac{Ax^2}{2!} + \frac{Bx^3}{3!} + aw_o\ \frac{x^4}{4!} - \frac{aw_o}{4!}\ (x-l)^4 \qquad x > l$$

Now when x = 2l, y = 0, y' = 0.

$$o = 2Al^2 + \frac{4}{3}\ Bl^3 + \frac{2}{3}\ aw_o\ l^4 - \frac{aw_o}{24}\ l^4$$

$$o = 2Al + 2Bl^2 + \frac{4}{3}\ aw_o\ l^3 - \frac{aw_o}{6}\ l^3$$

from which $\quad A = \frac{11}{48}\ l^2aw_o\ ,\ B = -\frac{13}{16}\ law_o$

$$\therefore\quad \frac{y}{aw_o} = \frac{11}{96}\ l^2x^2 - \frac{13}{96}\ lx^3 + \frac{1}{24}\ x^4 - \frac{1}{24}\ (x-l)^4\ U(x-l).$$

Application 3

273

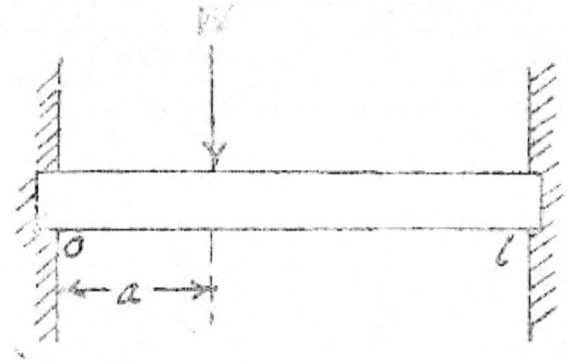

A beam of length l is clamped horizontally at its ends $x = 0$ and $x = l$ and carries a concentrated load W at $x = a$. Find the static deflection y at any point x.

$$EI \frac{d^4y}{dx^4} = W \, \delta(x-a)$$

$$\therefore \frac{d^4y}{dx^4} = \frac{W}{EI} \, \delta(x-a)$$

$$s^4\bar{y} - s^3 y(o) - s^2 y'(o) - s y''(o) - y'''(o) = \frac{W}{EI} e^{-as}$$

Now $y(o) = y'(o) = o$ and put $y''(o) = A$ and $y'''(o) = B$

$$\therefore \quad s^4\bar{y} = As + B + \frac{W}{EI} e^{-as}$$

$$\bar{y} = \frac{A}{s^3} + \frac{B}{s^4} + \frac{W}{EI} \frac{e^{-as}}{s^4}$$

$$y = A \frac{x^2}{2} + B \frac{x^3}{6} + \frac{W}{6EI} (x-a)^3 \, U(x-a)$$

Now at $x = l$ we have $y = o$, $y' = o$

$$o = Al^2 + \frac{B}{3} l^3 + \frac{W}{3EI} (l-a)^3$$

$$o = Al + \tfrac{1}{2} Bl^2 + \frac{W}{2EI} (l-a)^2$$

$$\therefore \quad A = \frac{wa (l-a)^2}{E \, l \, l^2} \; , \quad B = -\frac{w(l-a)^2 (l+2a)}{E \, l \, l^3}$$

274

$$\therefore \ EIy \ = \ \frac{wa(l-a)^2 \ x^2}{2l^2} \ - \ \frac{w(l-a)^2 \ (l+2a)x^3}{6l^3} \ + \ \frac{w(x-a)^3}{6}U(x-a).$$

Application 4

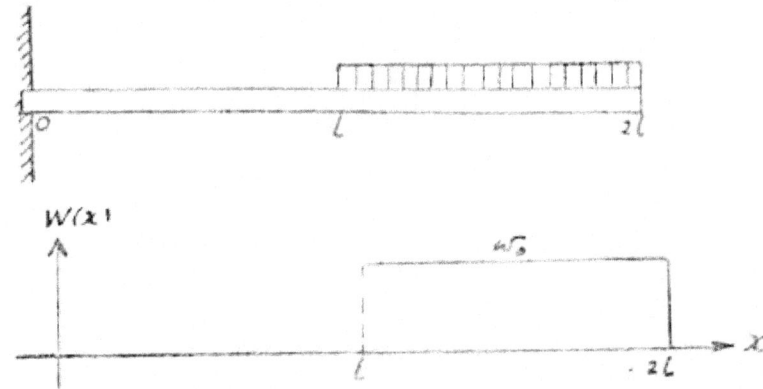

A cantilever of length $2l$ has its end x=0 built in while the end x=2l is free. It carries a load w(x) per unit length which is zero over $0<x<l$ and equal to a constant W_0 over $l<x<2l$. *Find* the static deflection at any point x of the cantilever.

$$EI \ \frac{d^4y}{dx^4} \ = \ w(x)$$

where $\quad w(x) \ = \ 0 \quad 0<x<l$

$$= \ w_0 \quad l<x<2l.$$

Extend the definition of w(x) so as to be equal to w o for all x>o

$$\frac{d^4y}{dx^4} \ = \ aw(x) \quad \text{where} \ a \ = \ \frac{1}{EI}$$

$$s^4\bar{y} \ - \ s^3y(o) \ - \ s^2y'(o) \ - \ sy''(o) \ - \ y'''(o) \ = \ \frac{aw_0}{s} \ e^{-ls}$$

Now $y(o) = o$, $y'(o) = o$ and put $y''(o) = A$, $y'''(o) = B$

$$\therefore \ s^4\bar{y} \ = \ As \ + \ B \ + \ \frac{aw_0}{s} \ e^{-ls}$$

$$\therefore \quad \bar{y} = \frac{A}{s^3} + \frac{B}{s^4} + \frac{aw_o}{s^5} \, e^{-ls}$$

$$y = A\frac{x^2}{2} + B\frac{x^3}{6} + \frac{aw_o}{24}(x-l)^4 U(x-l)$$

The constants A and B are determined *from* the conditions:
Y'''($2l$) = Y''($2l$) =0 since there is no bending shearing force at x = 2l

For s > l we have

$$Y''(x) = A + Bx + \tfrac{1}{2}aw_o(x-l)^2.$$

$$Y'''(x) = B + aw_o(x-l)$$

$$\therefore o = A + 2Bl + \tfrac{1}{2}aw_o l^2$$

$$o = B + aw_o l$$

$$\therefore A = \frac{3}{2}aw_o l^2 \,, \quad B = -aw_o l$$

$$\therefore y = aw_o\left[\frac{3}{4}l^2 x^2 - \frac{1}{6}lx^3 + \frac{1}{24}(x-l)^4 U(x-l)\right], \quad o \leqslant x \leqslant 2l$$

Application 5

A beam of length l is clamped horizontally at x=o and freely hinged at x= l It carries a load wx per unit length in $0<x<l/2$ and a load w(l -x) in $1/2<x<1$. Find the static deflection at any point x.

$$\frac{d^4y}{dx^4} = \frac{1}{EI}\left[wx - 2w\left(x-\frac{l}{2}\right)U\left(x-\frac{l}{2}\right)\right], \quad o<x<l$$

$$s^4\bar{y} - s^3y(o) - s^2y'(o) - sy''(o) - y'''(o)$$

$$= \frac{1}{EI}\left[\frac{w}{s^2} - \frac{2w}{s^3} e^{-\frac{l}{2}s}\right]$$

$y(o) = y'(o) = o$ and put $y''(o) = A$ and $y'''(o) = B$

$$s^4\bar{y} - As - B = \frac{1}{EI}\left[\frac{w}{s^2} - \frac{2w}{s^2} e^{-\frac{l}{2}s}\right]$$

$$\bar{y} = \frac{B}{s^4} + \frac{A}{s^3} + \frac{w}{EIs^6} - \frac{2w}{EIs^6} e^{-\frac{l}{2}s}$$

276

$$\therefore\ y\ =\ \frac{1}{6}\,Bx^3 + \tfrac{1}{2}\,Ax^2 + \frac{wx^5}{5!EI} - \frac{2w\left(x-\frac{l}{2}\right)^5}{5!\ EI}\,U\left(x-\frac{l}{2}\right)$$

Now we must have y = y" = 0 when x = l,

$$\therefore\ \frac{1}{6}\,Bl^3 + \tfrac{1}{2}\,Al^2 + \frac{15}{16}\,\frac{wl^5}{5!\ EI} = 0$$

$$Bl\ +\ A\ +\ \frac{3}{4}\,\frac{wl^3}{3!\ EI}\ =\ 0$$

Substituting the values of A and B from these two equations in the expression for y we get

$$EIy\ =\ \frac{5}{256}\,w\,x^2 l^3 - \frac{7}{256}\,w\,x^3 l^2 + \frac{1}{120}\,wx^5$$
$$-\ \frac{1}{60}\,w\,\left(x - \tfrac{1}{2}\,l\right)^5\,U\left(x - \frac{l}{2}\right).$$

Application 6

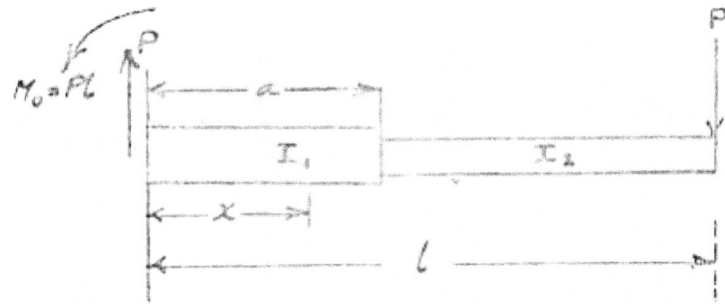

Determine the static deflection at any point of the cantilever beam shown above.

Solution:

We shall use the moment equation instead of the loading equation and the change in the stiffness is accounted for by a function of the form.

$$\frac{1}{EI}\,[\,1 + k\,U(x-a)\,]$$

The moment equation is

$$\frac{d^2y}{dx^2}\ =\ \frac{M}{EI_1}\,[\,1 - k\,U(x-a)\,]$$

where $\frac{1}{EI}$ over the section $0 \leqslant x \leqslant a$ is $\frac{1}{EI_1}$ and that over the section $x \geqslant a$ is

$$\frac{1}{EI_1}\,(1-k)\ =\ \frac{1}{EI_2}$$

from which $\quad k = 1 - \dfrac{I_1}{I_2}$

$\therefore \quad \dfrac{d^2y}{dx^2} = \dfrac{P(l-x)}{EI_1}\,[\,1 - kU(x-a)\,]$

Since $y(o) = y'(o) = o$, the subsidiary equation becomes

$$EI_1\; s^2\overline{y}(s) = \dfrac{Pl}{s} - \dfrac{P}{s^2} - \dfrac{Plke^{-as}}{s} + Pke^{-as}\left(\dfrac{a}{s} + \dfrac{1}{s^2}\right)$$

where $\quad L\left\{\,x\,U(x-a)\,\right\} = e^{-as}\left(\dfrac{a}{s} + \dfrac{1}{s^2}\right)$

Inverting we get

$$EI_1\;\; y(x) = Pl\,\dfrac{x^2}{2} - P\,\dfrac{x^3}{6}$$

$$- Pk\left[\dfrac{l(x-a)^2}{2} - \dfrac{(x-a)^3}{6} - \dfrac{a(x-a)^2}{2}\right]U(x-a)$$

Application 7

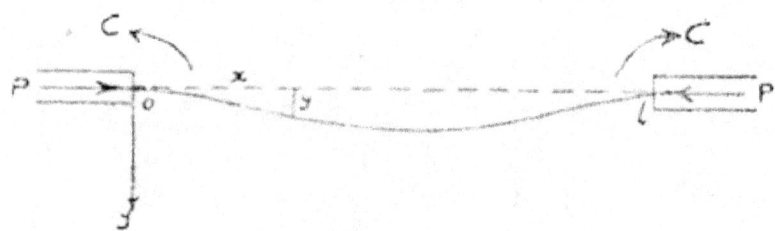

Find the critical loads for a strut clamped at both ends,

Solution:
Let the fixing couple at each end be C.

$$EI\,\dfrac{d^2y}{dx^2} = C - Py$$

$$EI\,[s^2\overline{y}(s) - sy(o) - y'(o)] = \dfrac{C}{s} - P\overline{y}(s)$$

$\therefore \quad (EIs^2 + P)\overline{y} = \dfrac{C}{s}$

$$\overline{y} = \dfrac{C}{s(EIs^2 + P)} = \dfrac{C}{EIs\left(s^2 + \dfrac{P}{EI}\right)}$$

$$= \frac{C}{EIs\,(s^2+n^2)} \quad \text{where } n^2 = \frac{P}{EI}$$

$$\therefore \quad y = \frac{C}{EI} \cdot \frac{1}{n^2} \left(1 - \cos nx \right)$$

$$\text{i.e.} \quad y = \frac{C}{P} \left(1 - \cos nx \right)$$

at $\quad x = l$, $y = 0 \quad \therefore 1 - \cos nl = 0 \quad \cos nl = 1$

$$\therefore \quad nl = 2\pi,\ 4\pi,\ \ldots$$

$$n = \frac{2\pi}{l}\ ,\ \frac{4\pi}{l}\ ,\ \ldots$$

First critical load when $n = \frac{2\pi}{l}$ is given by

$$P = n^2 EI = \frac{4\pi^2\,EI}{l^2}$$

Application 8

Critical loads for non-uniform columns.

279

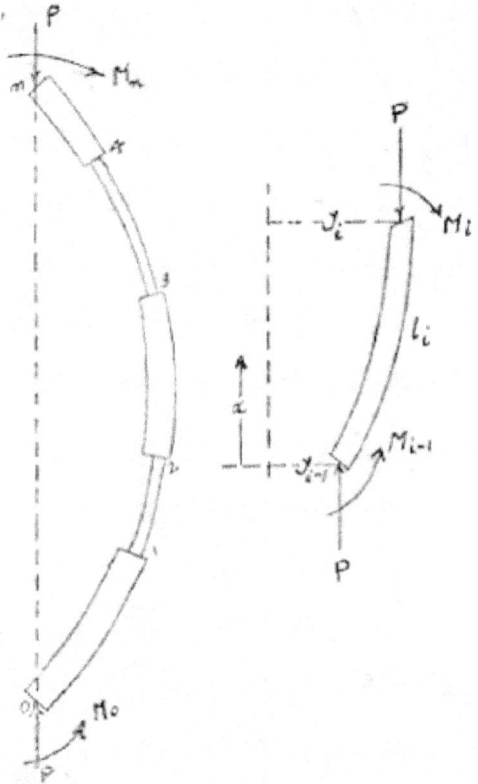

The operational method for treating columns of several sections is by shifting the origin to the end of each section.

Thus considering the i th part of the column shown in the above diagram. We have

$$E_i I_i \frac{d^2y}{dx^2} = -P(y - y_{i-1}) + M_{i-1}$$

Dividing by $E_i I_i$ and putting $\dfrac{P}{E_i I_i} = a_i^2$ we get

$$\frac{d^2y}{dx^2} = -a_i^2 (y - y_{i-1}) + \frac{M_{i-1}}{E_i I_i}$$

$$\therefore s^2 \overline{y(s)} - sy_{i-1} - y'_{i-1} = -a_i^2 y(s) + \frac{a_i^2 y_{i-1}}{s} + \frac{M_{i-1}}{E_i I_i s}$$

$$\therefore \quad (s^2 + a_i^2)\,\bar{y}(s) = s y_{i-1} + y'_{i-1} + \frac{a_i^2\, y_{i-1}}{s} + \frac{M_{i-1}}{E_i I_i s}$$

$$\therefore \quad \bar{y}(s) = \frac{y_{i-1}}{s} + \frac{y'_{i-1}}{s^2 + a_i^2} + \frac{M_{i-1}}{E_i I_i s(s^2 + a_i^2)}$$

Inverting we get

$$y(x) = y_{i-1} + \frac{y'_{i-1}}{a_i}\sin a_i x + \frac{M_{i-1}}{E_i I_i a_i^2}(1 - \cos a_i x)$$

Now in this equation and in the two equations obtained by differentiating it twice w.r.t x
put $\quad x = l_i$. so as to obtain the deflection, slope and bending moment at i in terms of the corresponding quantities at i —l.

$$y_i = y_{i-1} + \frac{y'_{i-1}}{a_i}\sin a_i l_i + \frac{M_{i-1}}{E_i I_i a_i^2}(1 - \cos a_i l_i)$$

$$y'_i = y'_{i-1}\cos a_i l_i + \frac{M_{i-1}}{E_i I_i a_i}\sin a_i l_i$$

$$M_i = -y'_{i-1} E_i I_i a_i \sin a_i l_i + M_{i-1}\cos a_i l_i$$

which may *be* expressed in matrix form as follows:

$$
\begin{bmatrix} y_i \\ y'_i \\ M_i \end{bmatrix}
=
\begin{bmatrix}
1 & \dfrac{1}{a_i}\sin a_i l_i & \dfrac{1}{E_i I_i a_i^2}(1 - \cos a_i l_i) \\[2ex]
0 & \cos a_i l & \dfrac{1}{E_i I_i a_i}\sin a_i l_i \\[2ex]
0 & -E_i I_i a_i \sin a_i l_i & \cos a_i l_i
\end{bmatrix}
\begin{bmatrix} y_{i-1} \\ y'_{i-1} \\ M_{i-1} \end{bmatrix}
$$

Y'

281

In the same way, by repeated application, we can

$$y_{i-1} \; , \; y'_{i-1} \; , \; M_{i-1}$$

in terms of $y_{i-2} \, , \, y'_{i-2} \, , \, M_{i-2}$ and so on.

Thus by multiplying the successive matrices we can obtain $y_n \, , \, y'_n \, , \, M_n$ in terms of $y_o \, , \, y'_o$ and M_o in the form

$$
\begin{bmatrix} y_n \\ y' \\ M_n \end{bmatrix}
=
\begin{bmatrix} P_{11} & P_{12} & P_{13} \\ 0 & P_{22} & P_{23} \\ 0 & P_{32} & P_{33} \end{bmatrix}
\begin{bmatrix} y_o \\ y'_o \\ M_o \end{bmatrix}
$$

and the critical loads can be obtained by substituting the boundary conditions in this equation. Thus for example in the case of a column whose ends are both built in we have

$$y_n = y_o = 0 \quad \text{and} \quad y'_n = y'_o = 0$$

and since

$$y_n = P_{11}y_o + P_{12}y'_o + P_{13}M_o$$

And

$$y'_n = P_{22}y'_o + P_{23}M_o$$

$$\therefore \; P_{13} = 0 \, , \; P_{23} = 0$$

If the column is hinged at both ends, then

$$y_n = y_o = 0 \quad \text{and} \quad M_n = M_o = 0$$

Now

$$y_n = P_{11}y_o + P_{12}y'_o + P_{13}M_o \; ,$$

$$M_n = P_{32}y'_o + P_{33}M_o$$

$$\therefore \qquad P_{12} = 0, \; P_{32} = 0$$

Application 9

A uniform beam with its ends x= 0 and x =./ built in carries a uniform load w per unit length. At x=a, there is an elastic support which provides a reaction equal to λ times the deflection. Determine the deflection at any point x of the beam.

Solution:

If R is the reaction at the support then

$$R = \lambda y(a)$$

$$EI \frac{d^4 y}{dx^4} = w - R\delta(x-a)$$

$$\therefore EI\ [s^4\bar{y} - s^3 y(o) - s^2 y'(o) - sy''(o) - y'''(o)\ ] = \frac{w}{s} - Re^{-as}$$

$$y(o) = y'(o) = o \text{ and put } y''(o) = A,\ y'''(o) = B, \text{ then}$$

$$\bar{y} = \frac{A}{s^3} + \frac{B}{s^4} + \frac{w}{EIs^5} - \frac{R}{EIs^4} e^{-as}$$

$$y = \tfrac{1}{2} Ax^2 + \frac{1}{6}Bx^3 + \frac{wx^4}{24EI} - \frac{R(x-a)^3}{6\ EI} U(x-a)$$

The unknowns A, B, R can be determined as follows :

Since y(l) = y'(1)= 0, then

$$o = \tfrac{1}{2} Al^2 + \frac{1}{6}Bl^3 + \frac{wl^4}{24\ EI} - \frac{R(l-a)^3}{6\ EI} \quad \ldots \quad \text{(i)}$$

$$o = Al + \tfrac{1}{2} Bl^2 + \frac{wl^3}{6EI} - \frac{R(l-a)^2}{2\ EI} \quad \ldots \quad \text{(ii)}$$

Since at x = a, y = R/λ then

$$\frac{R}{\lambda} = \tfrac{1}{2} Aa^2 + \frac{1}{6}Ba^3 + \frac{wa^4}{24EI} \quad \ldots \ldots \quad \text{(iii)}$$

Equations (i), (ii), (iii) determine A,B,R.

5.2. Exercises on Laplace Transform in practical applications

(1) A periodic e.m.f. E sin cut is applied at time t=o to an RC circuit in series. If the initial current and charge are zero, find the current I at any instant t.

(2) In problem (1) replace E sin wt by

$$E\ [U(t-t_o) - U(t-t_1)\ ]\ ,\quad t_1 > t_o > o.$$

(3) An e.m.f. E(t) is applied at time t o to an RC circuit in series. The initial current and charge are zero. Find the charge and current at any time t if
(i) E Eo a constant.
(ii)

$$E = E_o\ e^{-\alpha t}\ ,\quad \alpha > o.$$

(4) Same as in problem (3) but with E(t) $= E_o\delta(t)$ where $\delta(t)$ is the Dirac delta function.

(5) An electric circuit of an inductance L in series with a condenser of capacity C. At t=o an e.m.f. given by

$$E(t) = \frac{E_0 t}{T_0} \qquad o < t < T_0$$

$$= o \qquad t > T_0$$

is applied. Assuming zero initial current and charge, find the charge at any instant t.

(6) An e. .f. $E \cos(\omega t + a)$ is applied at time t = o to a series circuit of capacity C and inductance L. Find the current at any time t, assuming zero initial current and charge.

(7) In the circuit shown, calculate I_1, I_2, I_3 if E is constant and the initial currents and charges are zero.

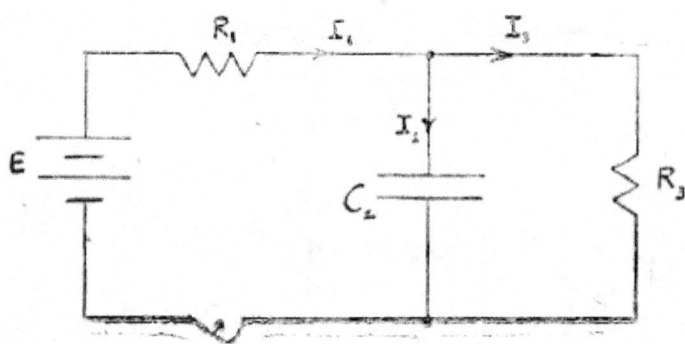

(8) In the circuit shown, the initial currents and charge are zero,

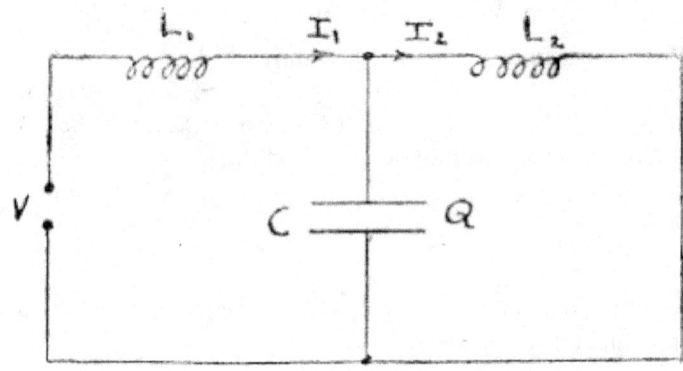

Find I_1 and I_2 when $V = E_0 \sin \omega t$.

(9) In the circuit shown, set up the equations for the determination of the currents I_1, I_2, I_3 and the charge Q_3. Transform the problem into algebraic form, assuming E constant and initial currents and charges to be zero.

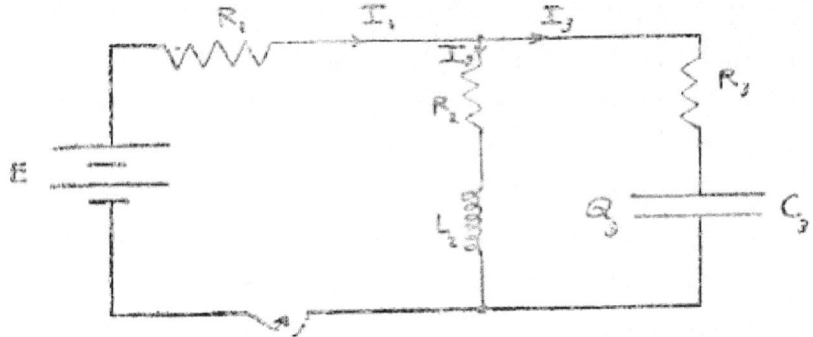

(10) A spring of stiffness k is placed on a smooth horizontal table. One end is fixed to a point 0 on the table and to the other end is attached a mass m. A force F(t), t >o acts on the mass. The differential equation of motion of the mass is

$$m\ddot{X} + kX = F(t)$$

Assuming $X(o) = a$, $\dot{X}(o) = o$ find $X(t)$ under the following conditions :

(i) $F(t) = F_o$ for $t > o$ where F_o is constant.

(ii) $F(t) = F_o e^{at}$ where $a > o$

(iii) $F(t) = F_o \sin \omega t$ where $\omega \neq \sqrt{\dfrac{k}{m}}$

(iv) $F(t) = F_o \sin \omega t$ where $\omega = \sqrt{\dfrac{k}{m}}$

(v) $F(t) = F_o U(t-T)$

(vi) $F(t) = F_o \delta(t-T)$.

(11) Same as in problem (10) but with initial conditions
$X(o) = \dot{X}(o) = o$. Find $X(t)$ when $F(t) = F_o\delta(t)$.

(12) A particle of mass m, at rest at the origin, is set in motion at t o by a blow of impulse P. Find its displacement at time t.

(13) A particle moves along a straight line such that its displacement x from a fixed point at time t is given by

$$\ddot{x} + 2\dot{x} + 4x = 20 \sin 4t$$

with initial conditions

$x(o) = \dot{x}(o) = o.$

Solve the equation and state which term in the result is the transient term and which is the steady state term. Find the amplitude and period of the steady state.

(14) Solve the equation of motion

$m\ddot{x} = F(t)$

of a particle with initial conditions

$x(o) = \dot{x}(o) = o$

where F(t) is given by

$$F(t) = \frac{2F_o}{T} t, \qquad o < t < \frac{T}{2}$$

$$= -\frac{2F_o}{T} (t-T), \frac{T}{2} < t < T$$

$$= o \qquad t > T$$

(15) A beam with its ends built in at x =o and x = l carries a uniform load W_o per unit length. Find the deflection at any point x.

(16) Work the same problem with the end x =0 built in and the end x = l hinged.

(17) A beam with its ends x = o and x = / hinged carries a load w(x) per unit length given by :

$$w(x) = o \qquad o < x < \frac{l}{4}$$

$$= w_o \qquad \frac{l}{4} < x < l,$$

find the static deflection at any point x.

(18) A cantilever beam with its end x=o built in and the end x =1 free carries a concentrated load P_o at x = 1/3, find the static deflection.

(19) A beam with its ends x=o and x=1 hinged carries a concentrated load Po at x = 1/4 . Find the deflection.

(20) A cantilever beam clamped at x = o and free at x= l carries a uniform load W_o per unit length. Find the deflection.

6. Using Laplace Transformation in solving Linear Partial Differential Equations.

The Laplace transformation can be used with advantage in solving linear partial differential equations. It is found that on applying this transformation the partial differential equation transforms to an ordinary differential equation, We illustrate by the following examples.

Example 1

Solve the partial differential equation

$$2x \frac{\partial Y}{\partial t} + \frac{\partial Y}{\partial x} = 2x \quad\dots\dots (1)$$

given that $Y(x,o) = 1$, $Y(0,t) = 1. .$, (2)
Writing equation (1) in the form

$$2xY_t (x,t) + Y_x (x,t) = 2x$$

and noticing that

$$L \left\{ Y_x (x,t) \right\} = \int_0^\infty e^{-st} \frac{\partial}{\partial x} Y(x,t) dt = \frac{\partial}{\partial x} \int_0^\infty e^{-st} Y(x,t)\ dt$$

$$= y_x (x,s)$$

we get on transforming (1) w.r.t. t

$$2x \left[sy(x,s) - Y(x,o) \right] + y_x (x,s) = \frac{2x}{s}$$

$$\therefore\ 2x \left[sy(x,s) - 1 \right] + y_x (x,s) = \frac{2x}{s}$$

$$\frac{dy}{dx} + 2xsy = 2x + \frac{2x}{s}$$

$$= 2x \left(1 + \frac{1}{s} \right)$$

This is a linear differential equation of the first order. The integrating factor is

$$e^{\int 2xsdx} = e^{sx^2}$$

$$\therefore\ ye^{sx^2} = \int 2x \left(1 + \frac{1}{s} \right) e^{sx^2} dx + C$$

$$= \frac{1}{s} \left(1 + \frac{1}{s} \right) e^{sx^2} + C$$

$$\text{i.e}\quad y(x,s) = \frac{1}{s} \left(1 + \frac{1}{s} \right) + Ce^{-sx^2} \quad\dots\dots (3)$$

Transforming (2) we get y(0,s) = 1/s

287

$$\frac{1}{8} = \frac{1}{s}\left(1+\frac{1}{s}\right) + C \quad \therefore \; C = -\frac{1}{s^2}$$

$$y(x,s) = \frac{1}{s}\left(1+\frac{1}{s}\right) - \frac{1}{s^2}\,e^{-sx^2}$$

$$= \frac{1}{s} + \frac{1}{s^2} - \frac{1}{s^2}\,e^{-x^2 s}$$

$$\therefore \; Y(x,t) = 1+t \qquad\qquad o \leqslant t \leqslant x^2$$

$$= 1+t-(t-x^2) \qquad t \searrow x^2$$

$$\text{i.e} \quad Y(x,t) = 1+t \qquad\qquad o \leqslant t \leqslant x^2$$

$$= 1+x^2 \qquad\qquad t \searrow x^2$$

Example 2

Find the solution of

$$\frac{\partial U}{\partial x} = 2\,\frac{\partial U}{\partial t} + U, \quad U(x,o) = 4e^{-2x}$$

which is bounded for x > o, t > o.

$$U_x(x,t) = 2U_t(x,t) + U(x,t)$$

$$\therefore \; u_x(x,s) = 2[su(x,s) - U(x,o)] + u(x,s)$$

$$= 2\left[su(x,s) - 4e^{-2x}\right] + u(x,s)$$

$$\frac{du}{dx} - (2s+1)u = -8e^{-2x}$$

which is a linear differential equation of the first order. Integrating factor is

$$e^{\displaystyle\int -(2s+1)dx} = e^{-(2s+1)x}$$

$$\therefore \; ue^{-(2s+1)x} = -8\int e^{-(2s+1)x}\,e^{-2x}\,dx + C$$

$$= -8 \int e^{-(2s+3)x} \, dx + C$$

$$= \frac{8}{2s+3} e^{-(2s+3)x} + C$$

$$\therefore \quad u(x,s) = \frac{8e^{-2x}}{2s+3} + Ce^{(2s+1)x}$$

Since U(x,t) must be bounded as $x \longrightarrow \infty$, we must have u(x,s) also bounded as $x \longrightarrow \infty$ $\therefore$ C = 0

$$u(x,s) = \frac{8}{2s+3} e^{-2x} = \frac{4}{s+\frac{3}{2}} e^{-2x}$$

$$\therefore \quad U(x,t) = 4e^{-\frac{3}{2}t} e^{-2x}$$

$$= 4e^{-\left(\frac{3}{2}t + 2x\right)}$$

Example 3
Solve the partial differential equation

$$xY_x(x,t) + Y_t(x,t) + Y(x,t) = x F(t)$$

with the boundary conditions
$$Y(x,0) = Y(0,t) = 0$$

Solution:

$$xy_x(x,s) + sy(x,s) - Y(x,0) + y(x,s) = xf(s)$$

$$\therefore \quad x\frac{dy}{dx} + sy + y = xf(s)$$

i.e $\quad \frac{dy}{dx} + \frac{s+1}{x} y = f(s)$

Integrating factor is

$$e^{\int \frac{s+1}{x} \, dx} = e^{(s+1)\log x} = x^{s+1}$$

289

$$\therefore \quad yx^{s+1} = \int f(s) \, x^{s+1} \, dx + C$$

$$= f(s) \, \frac{x^{s+2}}{s+2} + C$$

$$\therefore \quad y(x,s) = f(s) \, \frac{x}{s+2} + \frac{C}{x^{s+1}}$$

and since $Y(o,t) = o, \quad \therefore \quad y(o,s) = o, \quad \therefore \quad C = o$

$$y(x,s) = x \cdot \frac{1}{s+2} \, f(s)$$

$$\therefore \quad Y(x,t) = x \left(e^{-2t^*} F(t) \right)$$

$$= x \int_{o}^{t} e^{-2(t-\lambda)} \, F(\lambda) d\lambda$$

$$= xe^{-2t} \int_{o}^{t} e^{2\lambda} \, F(\lambda) d\lambda$$

Example 4

Solve the partial differential equation

$$U_{xx}(x,t) - 2U_{tx}(x,t) + U_{tt}(x,t) = o \quad (o < x < 1 \; , \; t > o)$$

with the boundary conditions

$$U(x,o) = U_t(x,o) = U(o,t) = o \; , \; U(1,t) = F(t) \quad t > o.$$

Solution:

$$L \left\{ U_{xx}(x,t) \right\} = \int_{o}^{\infty} e^{-st} \, \frac{\partial^2}{\partial x^2} \, U(x,t) dt$$

$$= \frac{\partial^2}{\partial x^2} \int_{o}^{\infty} e^{-st} U(x,t) dt = \frac{\partial^2}{\partial x^2} \, u(x,s) = u_{xx}(x,s)$$

$$\therefore \; u_{xx}(x,s) - 2\frac{\partial}{\partial x}[su(x,s)-U(x,o)]+s^2u(x,s)-sU(x,o)-U_t(x,o)=o$$

$$\therefore \; \frac{d^2u}{dx^2} - 2s\frac{du}{dx} + s^2u = o$$

$$(D^2 - 2sD + s^2)\,u = o \quad i.e \quad (D-s)^2u = o$$

$$\therefore \; u(x,s) = e^{sx}(A + Bx)$$

and since $U(o,t) = o \;\; \therefore\; u(o,s) = o$ and hence $A = o$

$$\therefore \; u(x,s) = Bxe^{xs}$$

Since $U(1,t) = F(t) \;\; \therefore\; u(1,s) = f(s)$

$$\therefore \; f(s) = Be^{s} \qquad i.e \; B = e^{-s}f(s)$$

$$\therefore \; u(x,s) = e^{-s}f(s)\,xe^{xs} \qquad i.e \; u(x,s) = xf(s)\,e^{-(1-x)s}$$

$$\therefore \; U(x,t) = o \qquad\qquad\qquad o<t< 1-x$$

$$= x\,F(t-1+x) \qquad\qquad t> 1-x$$

Example 5

Find a bounded solution of

$$\frac{\partial V}{\partial t} = k\frac{\partial^2 V}{\partial x^2} \qquad x>o\,,\, t>o$$

subject to the boundary conditions

$$V(o,t) = F(t)\,,\; V(x,o) = o$$

$$sv(x,s) - V(x,o) = kv_{xx}(x,s)$$

$$\frac{d^2v}{dx^2} - \frac{s}{k}\,v = o$$

$$\therefore \; v(x,s) = Ae^{\sqrt{s/k}\,x} + Be^{-\sqrt{s/k}\,x}$$

Since the solution is bounded, then $A = 0$

$$v(x,s) = Be^{-\sqrt{s/k}\,x}$$

$$v(x,s) = f(s)\,e^{-\sqrt{s/k}\,x}$$

Now $\quad L^{-1} \left\{ e^{-\alpha \sqrt{s}} \right\} = \dfrac{\alpha}{2\sqrt{\pi t^3}} \, e^{-\frac{\alpha^2}{4t}}$

$\therefore \quad V(x,t) = F(t)^{*} \dfrac{x}{2\sqrt{\pi k t^3}} \, e^{-\frac{x^2}{4kt}}$

$\qquad\qquad = \dfrac{x}{2\sqrt{\pi k}} \displaystyle\int_{0}^{t} \dfrac{F(\lambda)}{(t-\lambda)^{3/2}} \, e^{-\frac{x^2}{4k(t-\lambda)}} \, d\lambda$

6.1. Transverse vibrations of a stretched string under gravity.

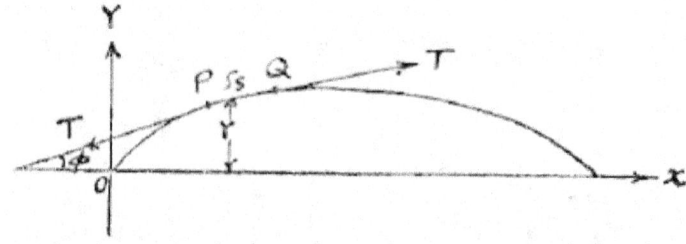

Let T be the tension in the string and m mass per unit length.
Consider an element PQ of the string of length δs.
Component of the tension at P resolved in the direction OY is

$$- T \sin \phi = - T \, \dfrac{\partial Y}{\partial s}$$

Component of the tension at Q resolved in the direction OY is

$$T \, \dfrac{\partial Y}{\partial s} + \dfrac{\partial}{\partial s} \left(T \dfrac{\partial Y}{\partial s} \right) \delta s$$

Resultant force due to the two tensions in the direction OY is

$$\dfrac{\partial}{\partial s} \left(T \dfrac{\partial Y}{\partial s} \right) \delta s = T \, \dfrac{\partial^2 Y}{\partial s^2} \, \delta s$$

$$\therefore \quad m \delta s \, \dfrac{\partial^2 Y}{\partial t^2} = T \, \dfrac{\partial^2 Y}{\partial s^2} \, \delta s \; - \; m \delta s \, g$$

$$\dfrac{\partial^2 Y}{\partial t^2} = \dfrac{T}{m} \, \dfrac{\partial^2 Y}{\partial s^2} \; - \; g. \qquad \text{Put} \quad \dfrac{T}{m} = a^2$$

$$\therefore \quad \dfrac{\partial^2 Y}{\partial t^2} = a^2 \, \dfrac{\partial^2 Y}{\partial s^2} \; - \; g$$

Assuming small lateral displacements and gradients we can replace

292

$\dfrac{\partial^2 Y}{\partial s^2}$ by $\dfrac{\partial^2 Y}{\partial x^2}$

$\therefore \quad \dfrac{\partial^2 Y}{\partial t^2} = a^2 \dfrac{\partial^2 Y}{\partial x^2} - g$

i.e $\quad Y_{tt}(x,t) = a^2 Y_{xx}(x,t) - g$

Example 1.

A semi - infinite stretched string of negligible weight has its distant end fixed while the end $x = 0$ is initially at the origin and moves along the Y — axis such that $Y(t) = C \sin \omega t$, $t > 0$. The string is initially along the x-axis with no initial velocity. When the end $x = 0$ starts to move find the shape of the string at any subsequent instant.

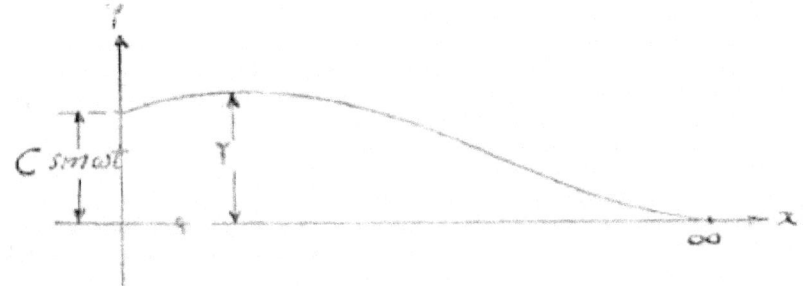

Differential equation of motion of the string is

$$Y_{tt}(x,t) = a^2 Y_{xx}(x,t) \quad . \quad . \quad . \quad (1)$$

At $t=0$ displacements and velocities are zero.

$$Y(x,0) = 0, \quad Y_t(x,0) = 0 \quad . \quad . \quad . \quad (2)$$

At $x=0$, $Y(0,t) = C \sin \omega t \quad . \quad . \quad . \quad . \quad (3)$

At $x = \infty$, $\quad \lim_{x \to \infty} Y(x,t) = 0 \quad . \quad . \quad . \quad (4)$

$$\therefore \quad y(x,s) = A e^{\frac{s}{a} x} + B e^{-\frac{s}{a} x}$$

Transforming (1) w.r.t. t we get

$$s^2 y(x,s) - s Y(x,0) - Y_t(x,0) = a^2 y_{xx}(x,s)$$

$$\therefore \ s^2 y = a^2 \frac{d^2 y}{dx^2} \quad \text{i.e} \quad \frac{d^2 y}{dx^2} - \frac{s^2}{a^2} y = 0$$

$$\therefore \ y(x,s) = A e^{\frac{s}{a} x} + B e^{-\frac{s}{a} x}$$

Equation (4) transforms into

$$\lim_{x \to \infty} y(x,s) = 0 \quad \therefore \ A = 0$$

$$y(x,s) = B e^{-\frac{s}{a} x}$$

Equation (3) transforms into $y(o,s) = \dfrac{C\omega}{s^2 + \omega^2}$

$$\therefore \quad B = \frac{C\omega}{s^2 + \omega^2}$$

$$y(x,s) = \frac{C\omega}{s^2 + \omega^2} . e^{-\frac{s}{a} x}$$

$$Y(x,t) = 0 \qquad\qquad 0 < t < \frac{x}{a}$$

$$= C \sin \omega \left(t - \frac{x}{a} \right) \quad t > \frac{x}{a}$$

The motion may be interpreted as follows:

A point at distance x from the origin remains at rest until a time t = x/a elapses when the motion at the origin is transferred to the point. The time t = x/a is the time necessary for a wave traveling with velocity a to describe a distance x.

Example 2

A semi - infinite string has its end x = o fixed while the distant and is looped around a vertical support that cannot exert any vertical force upon the string i.e.,

$$T \frac{\partial Y}{\partial x} = 0$$

at that end. The string is initially supported along the x – axis and at time t.= o the support is removed and the string moves downwards under the action of gravity. Find the shape of the string at any instant.

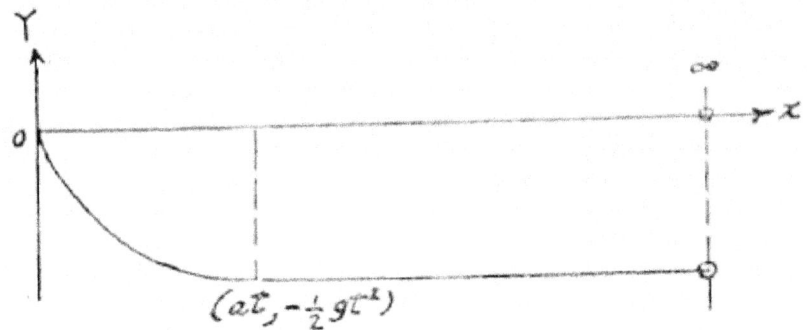

$$\left(at, -\frac{1}{2}gt^2\right)$$

Solution:

$$Y_{tt}(x,t) = a^2 Y_{xx}(x,t) - g \quad . \quad . \quad . \quad (1)$$

At time t = 0, displacements and velocities at all points of the string are zero.

$$\therefore \quad Y(x,0) = 0, \ Y_t(x,o) = 0 \quad . \quad . \quad . \quad (2)$$

At the end x= 0 , Y = 0 for all t, i.e.,

$$Y(0,t) = 0 \ldots\ldots\ldots\ldots\ldots\ldots\ldots\ldots\ldots\ldots \quad (3)$$

At the distant end x = ∞ , the vertical component of the tension is zero.

$$\therefore \quad \lim_{x \to \infty} Y_x(x,t) = 0 \quad . \quad . \quad . \quad (4)$$

Transforming (1) w.r.t. t

$$s^2 y(x,s) = a^2 y_{xx}(x,s) - \frac{g}{s}$$

$$\therefore \quad \frac{d^2 y}{dx^2} - \frac{s^2}{a^2} y = \frac{g}{a^2 s}$$

$$y(x,s) = Ae^{\frac{s}{a}x} + Be^{-\frac{s}{a}x} - \frac{g}{s^3}$$

Equation (4) transforms into

$$\lim_{x \to \infty} y_x(x,s) = 0 \quad \therefore \quad A = 0$$

$$\therefore \quad y(x,s) = Be^{-\frac{s}{a}x} - \frac{g}{s^3}$$

295

Again equation (g) transforms into

$$y(0,s) = 0 \quad \therefore \quad B = \frac{g}{s^3}$$

$$y(x,s) = \frac{g}{s^3} \left(e^{-\frac{s}{a}x} - 1 \right)$$

$$= -\frac{g}{2} \left(\frac{2}{s^3} - \frac{2}{s^3} e^{-\frac{x}{a}s} \right)$$

$$\therefore \quad Y(x,t) = -\tfrac{1}{2} gt^2 \qquad\qquad 0 < t < \frac{x}{a}$$

$$= -\tfrac{1}{2} gt^2 + \tfrac{1}{2} g \left(t - \frac{x}{a} \right)^2 \qquad t > \frac{x}{a}$$

i.e $\quad Y(x,t) = -\tfrac{1}{2} gt^2 \qquad\qquad 0 < t < \frac{x}{a}$

$$= -\frac{g}{2a^2} (2axt - x^2) \qquad t > \frac{x}{a}$$

$$\therefore \quad Y(x,t) = -\frac{g}{2a^2} (2axt - x^2) \qquad x < at$$

$$= -\tfrac{1}{2} gt^2 \qquad\qquad x > at$$

Considering the $f$irst equation

$$x^2 - 2axt = \frac{2a^2 Y}{g}$$

$$\therefore \quad (x - at)^2 = \frac{2a^2 Y}{g} + a^2 t^2$$

i.e $\quad (x - at)^2 = \frac{2a^2}{g} (Y + \tfrac{1}{2} gt^2)$

which is a parabola vertex

$$\left(at , -\tfrac{1}{2} gt^2 \right)$$

and whose axis is parallel to ()Y.

In the instantaneous position of the string, we notice that, at any time t, all elements of the string to the right of the point x at move like freely falling bodies.

Example 3

296

A semi infinite stretched string has its distant end fixed on the x — axis while the end x = 0 is looped around the Y - axis which exerts, no vertical force on the loop. The string is initially at rest in the position $Y = e^{-x}$ released from this position with no external forces acting on it. Find the displacement of the string at any point x and at any instant t.

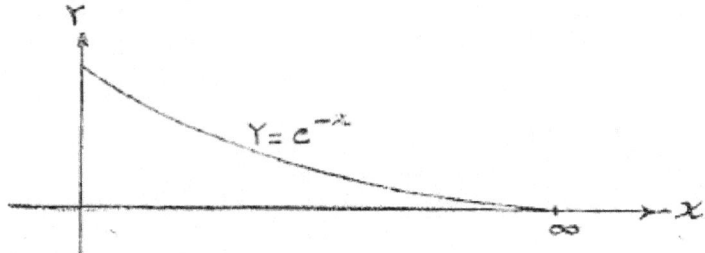

Solution:

The equation of motion of the string is

$$Y_{tt}(x,t) = a^2 Y_{xx}(x,t)$$

Boundary conditions are

$$Y(x,0) = e^{-x} \,,\; Y_t(x,0) = 0 \,,\; Y_x(0,t) = 0 \,,\quad \lim_{x \to \infty} Y(x,t) = 0.$$

$$s^2 y(x,s) - s\, Y(x,0) - Y_t(x,0) = a^2 y_{xx}(x,s)$$

$$\therefore \quad s^2 y(x,s) - se^{-x} = a^2 y_{xx}(x,s)$$

$$\frac{d^2 y}{dx^2} - \frac{s^2}{a^2} y = -\frac{s}{a^2} e^{-x}$$

$$\therefore \quad y(x,s) = Ae^{\frac{s}{a}x} + Be^{-\frac{s}{a}x} + \frac{1}{D^2 - \frac{s^2}{a^2}}\left(-\frac{s}{a^2} e^{-x}\right)$$

$$y(x,s) = Ae^{\frac{s}{a}x} + Be^{-\frac{s}{a}x} - \frac{s}{a^2 - s^2}e^{-x}$$

$$\lim_{x \to \infty} y(x,s) = 0 \qquad \therefore \quad A = 0$$

$$y(x,s) = Be^{-\frac{s}{n}x} - \frac{s}{a^2 - s^2}e^{-x}$$

$$y_x(0,s) = 0 \text{ and } y_x(x,s) = -\frac{s}{a}Be^{-\frac{s}{a}x} + \frac{s}{a^2 - s^2}e^{-x}$$

$$\therefore \quad y_x(0,s) = -\frac{s}{a}B + \frac{s}{a^2 - s^2} = 0 \therefore B = \frac{a}{a^2 - s^2}$$

$$\therefore \quad y(x,s) = -\frac{a}{s^2 - a^2}e^{-\frac{s}{a}x} + \frac{s}{s^2 - a^2}e^{-x}$$

$$\therefore \quad Y(x,t) = e^{-x} \cosh at \qquad\qquad 0 < t < \frac{x}{a}$$

$$= e^{-x} \cosh at - \sinh a\left(t - \frac{x}{a}\right) \qquad t > \frac{x}{a}$$

i.e $\quad Y(x,t) = e^{-x} \cosh at + \sinh(x - at) \qquad x < at$

$$= e^{-x} \cosh at \qquad\qquad x > at.$$

Example 4

A string is stretched between the two fixed points x = 0 and x = 1. The string has initially the shape

$$Y = b \sin \frac{\pi x}{l}$$

and is released from rest at t = 0 in that position. Find the shape of the string at any subsequent instant.

Solution:

The differential equation of motion of the string is

$$Y_{tt}(x,t) = a^2 Y_{xx}(x,t) \qquad\qquad 0 < x < l, \ t > 0.$$

The boundary conditions are

$$Y(x,o) = b \sin \frac{\pi x}{l} \ , \ Y_t(x,o) = o \ , \ Y(o,t) = Y(l,t) = o$$

$$s^2 y(x,s) - sY(x,o) - Y_t(x,o) = a^2 y_{xx}(x,s)$$

$$\therefore \quad s^2 y - bs \sin \frac{\pi x}{l} = a^2 \frac{d^2 y}{dx^2}$$

i e $\quad \dfrac{d^2 y}{dx^2} - \dfrac{s^2}{a^2} y = -\dfrac{bs}{a^2} \sin \dfrac{\pi x}{l}$

$$y(x,s) = A \cosh \frac{s}{a} x + B \sinh \frac{s}{a} x + \frac{1}{D^2 - \dfrac{s^2}{a^2}}\left(-\frac{bs}{a^2} \sin \frac{\pi x}{l}\right)$$

$$= A \cosh \frac{s}{a} x + B \sinh \frac{s}{a} x + \frac{1}{-\dfrac{\pi^2}{l^2} - \dfrac{s^2}{a^2}}\left(-\frac{bs}{a^2} \sin \frac{\pi x}{l}\right)$$

$$= A \cosh \frac{s}{a} x + B \sinh \frac{s}{a} x + \frac{bl^2 s}{\pi^2 a^2 + l^2 s^2} \sin \frac{\pi x}{l}$$

Since $Y(o,t) = o$ ∴ $y(o,s) = o$ ∴ $A = o$

$$y(x,s) = B \sinh \frac{s}{a} x + \frac{bl^2 s}{\pi^2 a^2 + l^2 s^2} \sin \frac{\pi x}{l}$$

Since $Y(l,t) = o$ ∴ $y(l,s) = o$

$$o = B \sinh \frac{s}{a} l \qquad ∴ \quad B = o$$

Hence $y(x,s) = \dfrac{bl^2 s}{\pi^2 a^2 + l^2 s^2} \sin \dfrac{\pi x}{l}$

$$= bl^2 \sin \frac{\pi x}{l} \cdot \frac{1}{l^2} \cdot \frac{s}{s^2 + \dfrac{\pi^2 a^2}{l^2}}$$

$$= b \sin \frac{\pi x}{l} \cdot \frac{s}{s^2 + \dfrac{\pi^2 a^2}{l^2}}$$

∴ $Y(x,t) = b \sin \dfrac{\pi x}{l} \cos \dfrac{\pi a}{l} t$

6.2. Longitudinal vibrations of bars

Let one end O of an elastic bar be fixed and let A be the cross sectional area of the bar, p its density and E its Young's modulus.

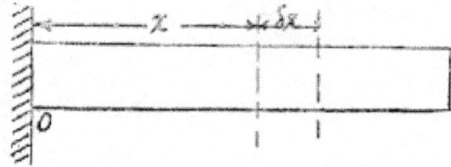

Consider an element of length δx of the bar, Its mass is $\rho A \, \delta x$. Suppose that the length x extends a distance Y then a length δx extends a distance

$$\frac{\partial Y}{\partial x} \delta x$$

The strain in the element is

300

$\left(\frac{\partial Y}{\partial x} \delta x\right) / \delta x$ i.e $\frac{\partial Y}{\partial x}$

$\text{Stress} = E \frac{\partial Y}{\partial x}$. $\text{Tension} = EA \frac{\partial Y}{\partial x}$

This is at a distance x. Tension at

$x + \delta x$ is $EA\frac{\partial Y}{\partial x} + \frac{\partial}{\partial x}\left(EA\frac{\partial Y}{\partial x}\right)\delta x.$

$\therefore$ Resultant tension in element $= EA \frac{\partial^2 Y}{\partial x^2} \delta x$. This should

be equal to $\rho A \, \delta x \, \frac{\partial^2 Y}{\partial t^2}$.

$\therefore \quad \rho A \, \delta x \, \frac{\partial^2 Y}{\partial t^2} = EA \frac{\partial^2 Y}{\partial x^2} \delta x$

i.e $\quad \frac{\partial^2 Y}{\partial t^2} = \frac{E}{\rho}\frac{\partial^2 Y}{\partial x^2}$. Put $\frac{E}{\rho} = a^2$

$\therefore \quad Y_{tt}(x,t) = a^2 \, Y_{xx}(x,t).$

Example 1

$x = 0$ $x = l$ F_o

An elastic bar of length /, has its end x=o fixed while the other end x=1 is acted upon by a constant force Fo per unit area parallel to the bar. The bar is initially unstrained and at rest. Find the longitudinal displacement of the end x = / at any instant.

$$Y_{tt}(x,t) = a^2 Y_{xx}(x,t) \quad . \quad . \quad . \quad . \quad (1)$$

At t=o $\quad Y(x,o) = o, \; Y_t(x,o) = o \quad . \quad . \quad (2)$

At x=o $\quad Y(o,t) = o \quad . \quad . \quad . \quad . \quad . \quad . \quad (3)$

At x=/ , stress $= Fo$

$\therefore \quad EY_x(l,t) = Fo \quad . \quad . \quad . \quad . \quad (4)$

Transforming (1) we get

$$s^2 y(x,s) - sY(x,0) - Y_t(x,0) = a^2 y_{xx}(x,s)$$

$$\therefore \quad s^2 y(x,s) = a^2 y_{xx}(x,s)$$

$$\frac{d^2 y}{dx^2} - \frac{s^2}{a^2} y = 0$$

$$y(x,s) = A \cosh \frac{s}{a} x + B \sinh \frac{s}{a} x$$

Transforming (3) we get

$$y(0,s) = 0 \quad \therefore \quad A = 0$$

$$\therefore \quad y(x,s) = B \sinh \frac{s}{a} x$$

Transforming (4) we get

$$Ey_x(l,s) = \frac{Fo}{s}$$

$$E.B \frac{s}{a} \cosh \frac{s}{a} l = \frac{F}{s}$$

$$\therefore B = \frac{aF_o}{E} \cdot \frac{1}{s^2} \operatorname{sech} \frac{s}{a} l$$

$$y(x,s) = \frac{aF_o}{E} \cdot \frac{1}{s^2} \operatorname{sech} \frac{s}{a} l \sinh \frac{s}{a} x$$

$$\therefore y(l,s) = \frac{aF_o}{E} \frac{1}{s^2} \tanh \frac{s}{a} l$$

Now we already know that the Laplace transform of the triangular wave H(c,t) defined by

$$H(c,t) = t \qquad 0 < t < c$$

$$= 2c - t \qquad c < t < 2c$$

$$H(c,t + 2c) = H(c,t)$$

is given by

$$L\left\{ H(c,t) \right\} = \frac{1}{s^2} \tanh \frac{cs}{2}$$

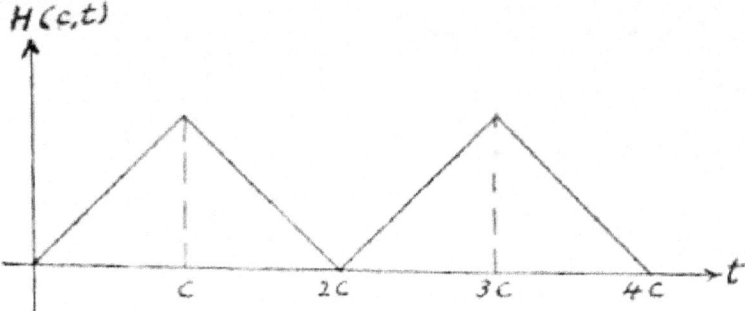

Hence $Y(l,t) = \dfrac{aF_o}{E} H\left(\dfrac{2l}{a}, t\right)$

Example 2

A uniform vertical rod of mass m and length l is clamped at the end x = o and is mass loaded by a mass M at its other end x = l. Investigate the behavior of the system to an arbitrary excitation F(t) applied to the mass M.

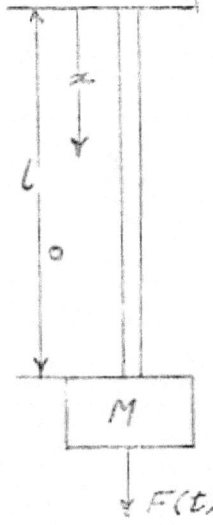

Solution:

The partial differential equation satisfied by the displacement Y of an element at distance x from the top is

$$\dfrac{\partial^2 Y}{\partial t^2} = a^2 \dfrac{\partial^2 Y}{\partial x^2} \quad \text{where } a = \sqrt{\dfrac{E}{\rho}}$$

i.e $Y_{tt}(x,t) = a^2 Y_{xx}(x,t)$

and since the system is initially at rest, its transform is

$$s^2 y(x,s) = a^2 y_{xx}(x,s)$$

i.e $$\frac{d^2y}{dx^2} - \frac{s^2}{a^2} y = 0$$

$$\therefore \quad y(x,s) = B \cosh \frac{s}{a} x + C \sinh \frac{s}{a} x$$

and since $Y(o,t) = o$ $\therefore$ $y(o,s) = o$ $\therefore$ $B = o$ and hence

$$y(x,s) = C \sinh \frac{s}{a} x$$

The constant C can be determined from the boundary condition at x = l.

$$M \left[\frac{\partial^2 Y}{\partial t^2} \right]_{x=l} = F(t) - AE \left[\frac{\partial Y}{\partial x} \right]_{x=l}$$

where A is the cross sectional area of the rod.

$$\therefore \quad Ms^2 y(l,s) = f(s) - AE y_x(l,s)$$

Now, since $y(x,s) = C \sinh \frac{s}{a} x$

$$\therefore \quad y(l,s) = C \sinh \frac{s}{a} l$$

and $y_x(l,s) = C \frac{s}{a} \cosh \frac{s}{a} l$

$$\therefore \quad Ms^2 \, C \sinh \frac{s}{a} l = f(s) - AEC \frac{s}{a} \cosh \frac{s}{a} l$$

$$C \left(Ms^2 \sinh \frac{s}{a} l + A E \frac{s}{a} \cosh \frac{s}{a} l \right) = f(s)$$

$$\therefore \quad C = \frac{f(s)}{Ms^2 \sinh \frac{s}{a} l + AE \frac{s}{a} \cosh \frac{s}{a} l}$$

$$\therefore \; y(x,s) = \frac{f(s) \sinh \frac{s}{a} x}{Ms^2 \sinh \frac{sl}{a} + AE \frac{s}{a} \cosh \frac{sl}{a}}$$

$$\therefore \; Y(x,t) = \frac{1}{2\pi i} \int_{\gamma - i\infty}^{\gamma + i\infty} \frac{e^{st} f(s) \sinh \frac{s}{a} x \; ds}{Ms^2 \sinh \frac{sl}{a} + AE \frac{s}{a} \cosh \frac{sl}{a}}$$

6.3. Partial differential equations of transmission lines

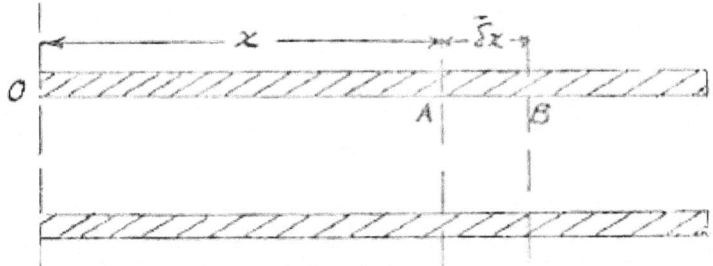

Let L,C,R,G be respectively the inductance, capacitance, resistance find dielectric conductance or leakance, all per unit length of a transmission line. Consider an element of length §x of this line. Let the potential at A be V, then the potential at B is

$$V + \frac{\partial V}{\partial x} \delta x$$

The drop of potential is

$$- \frac{\partial V}{\partial x} \delta x.$$

This is partly due to resistance $R\delta x$ and partly due to inductance $L\delta x$

$$\therefore \; IR\delta x + L\delta x \frac{\partial I}{\partial t} = - \frac{\partial V}{\partial x} \delta x.$$

i.e $\left(L \frac{\partial}{\partial t} + R \right) I = - \frac{\partial V}{\partial x}$ . . . (1)

Next consider an interval of time δt Charge of electricity entering the element through section A in this interval is $I\delta t$. Charge leaving the element out of the section B in the same interval is

$$\left(I + \frac{\partial I}{\partial x} \delta x \right) \delta t.$$

Hence net outflow of charge from these two sections is

$$\frac{\partial I}{\partial x}\, \delta x \delta t.$$

To this we should add the charge leaking from core to sheath namely

$$G \delta x V \delta t.$$

Hence the total outflow of charge from element is

$$\frac{\partial I}{\partial x}\, \delta x \delta t \;+\; G \delta x V \delta t$$

This will cause a drop of potential

$$-\frac{\partial V}{\partial t}\, \delta t$$

$$\therefore \quad \frac{\dfrac{\partial I}{\partial x}\, \delta x \delta t \;+\; G \delta x V \delta t}{C \delta x} \;=\; -\frac{\partial V}{\partial t}\, \delta t$$

from which we get

$$\left(C\, \frac{\partial}{\partial t} + G \right) V \;=\; -\frac{\partial I}{\partial x} \qquad . \quad . \quad (2)$$

Now operating on both sides of this equation *by*

$$L\, \frac{\partial}{\partial t} + R$$

we get

$$\left(L\, \frac{\partial}{\partial t} + R \right)\left(C\, \frac{\partial}{\partial t} + G \right) V \;=\; -\left(L\, \frac{\partial}{\partial t} + R \right)\frac{\partial I}{\partial x}$$

$$=\; -\frac{\partial}{\partial x}\left(L\, \frac{\partial}{\partial t} + R \right) I$$

$$=\; -\frac{\partial}{\partial x}\left(-\frac{\partial V}{\partial x} \right) = \frac{\partial^2 V}{\partial x^2}$$

$$\therefore \quad CL\, \frac{\partial^2 V}{\partial t^2} + (CR + LG)\frac{\partial V}{\partial t} + RGV \;=\; \frac{\partial^2 V}{\partial x}$$

This is the partial differential equation satisfied by V and we have a similar equation satisfied by I namely

$$CL \frac{\partial^2 I}{\partial t^2} + (CR + LG) \frac{\partial I}{\partial t} + RGI = \frac{\partial^2 I}{\partial x^2} \, .$$

Now let us apply the Laplace transformation to the above equations.

Write equations (1) and (2) in the form:

$$LI_t(x,t) + RI(x,t) = -V_x(x,t)$$

$$CV_t(x,t) + GV(x,t) = -I_x(x,t)$$

Transforming w.r.t. t we get:

$$L \{ si(x,s) - I(x,o) \} + Ri(x,s) = -v_x(x,s)$$

$$C \{ sv(x,s) - V(x,o) \} + Gv(x,s) = -i_x(x,s)$$

$$\therefore (Ls+R) \, i(x,s) = LI(x,o) - v_x(x,s) \quad \ldots \ldots (3)$$

$$(Cs+G) \, v(x,s) = CV(x,o) - i_x(x,s) \quad \ldots \ldots (4)$$

Eliminating i(x,s) between (3) and (4) by differentiating (3) partially w.r.t. x and substituting for i(x,s) from (4) we get

$$\frac{d^2v}{dx^2} - (Ls+R)(Cs+G) \, v = LI_x(x,o) - C(Ls+R)V(x,o)$$

Putting $q^2 = (Ls+R)(Cs+G)$ we get

$$\frac{d^2v}{dx^2} - q^2v = LI_x(x,o) - C(Ls+R) \, V(x,o)$$

Now suppose we have a semi— infinite line $x \geq o$ with the following conditions:

(i) Initial currents and charges are zero.

(ii) $\lim\limits_{x \to \infty} V(x,t) = o$

(iii) $V(o,t) = F(t)$

Hence the differential equation in v(x,s) reduces to

$$\frac{d^2v}{dx^2} - q^2v = o \quad \text{where } q^2 = (Ls+R)(Cs+G)$$

and we have to solve his equation subject to the two conditions

$$\lim_{x \to \infty} v(x,s) = 0 \quad \text{and} \quad v(o,s) = f(s)$$

$$v(x,s) = Ae^{qx} + Be^{-qx}$$

and since $v(o,s) = f(s)$ then $f(s) = B$

Hence

$$v(x,s) = f(s)e^{-qx}$$

i.e $\quad v(x,s) = f(s)e^{-\sqrt{(Ls+R)(Cs+G)} \; x}$

The inverse transform in the general case is a complicated one. The following special but important cases will now be investigated.

Case I.
The lossless line $R = 0$, $G = 0$

Here $\quad v(x,s) = f(s)e^{-\sqrt{CL} \, sx}$. Put $u = \dfrac{1}{\sqrt{CL}}$

$$\therefore \quad v(x,s) = f(s) \, e^{-\frac{x}{u}s}$$

$$\therefore \quad V(x,t) = 0 \qquad 0 < t < \frac{x}{u}$$

$$= F\left(t - \frac{x}{u}\right) \qquad t > \frac{x}{u}$$

This means that a point at distance x from the origin remains at zero potential until a time $t = x/u$ elapses when the voltage $F(t)$ at $x = 0$ is transferred to the point. The time x/u is the time necessary for a wave traveling with velocity u to describe the distance x.

Case II.

The Heaviside's distortionless line $CR = LG$.

$$\text{Put} \quad \frac{R}{L} = \frac{G}{C} = \lambda \qquad \therefore \quad R = \lambda L \, , \; G = \lambda C$$

$$\therefore \quad q^2 = (Ls+R)(Cs+G) = (Ls+\lambda L)(Cs+\lambda C) = CL(s+\lambda)^2$$

$$\therefore \quad q = \sqrt{CL}\,(s+\lambda) = \frac{s+\lambda}{u} \quad \text{where} \quad u = \frac{1}{\sqrt{CL}}$$

308

$$\therefore \quad v(x,s) = f(s)\, e^{-\dfrac{s+\lambda}{u}x}$$

$$\text{i.e} \quad v(x,s) = f(s)\, e^{-\dfrac{\lambda x}{u}}\, e^{-\dfrac{x}{u}s}$$

$$\therefore \quad V(x,t) = 0 \qquad\qquad 0 < t < \frac{x}{u}$$

$$= e^{-\dfrac{\lambda x}{u}}\, F\left(t - \frac{x}{u}\right) \qquad t > \frac{x}{u}$$

Hence we have a solution similar to the previous case but with attenuation as we move along the line due to the existence of the factor

$$e^{-\dfrac{\lambda x}{u}}$$

Case III.
The ideal submarine cable L=0, G=0.

$$\text{Since } q^2 = (Ls+R)(Cs+G) \quad \text{then} \quad q^2 = RCs$$

$$\therefore \quad q = \sqrt{RCs} = \sqrt{\frac{s}{k}} \qquad \text{where} \quad k = \frac{1}{RC}$$

$$\therefore \quad v(x,s) = f(s)\, e^{-\sqrt{s/k}\, x}$$

Suppose that F(t) E where E is constant, then f(s) = E / s

$$\therefore \quad v(x,s) = \frac{E}{s}\, e^{-\sqrt{s/k}\, x}$$

$$\text{Now } L^{-1}\left\{ \frac{1}{s}\, e^{-\sqrt{s/k}\, x} \right\} = \operatorname{erfc}\frac{x}{2\sqrt{kt}}$$

$$\therefore \quad V(x,t) = E\,\operatorname{erfc}\left(\frac{x}{2\sqrt{kt}}\right)$$

This problem, as we shall see, has an analogous problem in linear heat flow.

6.4. Conduction of heat

Let p = density of conducting medium, c its specific heat and K its thermal conductivity. Consider an infinitesimal rectangular parallelepiped whose centre P is the point (x,y,z) and whose faces are parallel to the coordinate planes and let the lengths of its sides be δx, δy, δz.

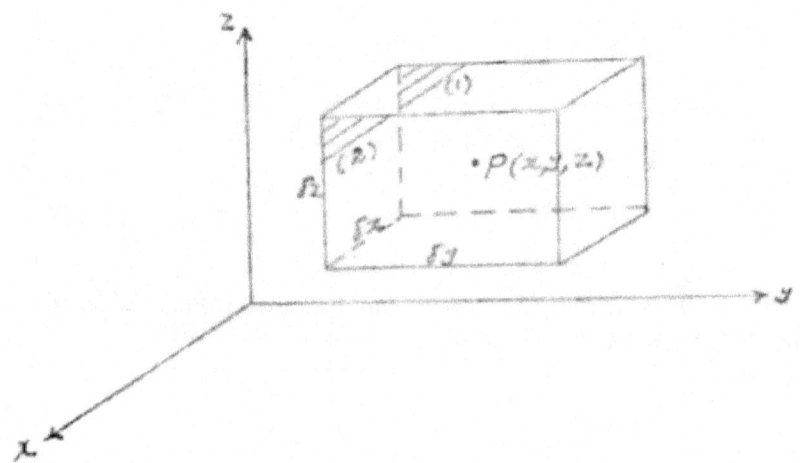

The quantity of heat entering the element in time δt through the face (1) is

$$- K \ \delta y \delta z \left[\frac{\partial U}{\partial x} - \frac{\partial^2 U}{\partial x^2} \ \frac{\delta x}{2} \right] \delta t$$

The quantity of heal leaving the element in time δt from face (2) is

$$- K \ \delta y \delta z \left[\frac{\partial U}{\partial x} + \frac{\partial^2 U}{\partial x^2} \ \frac{\delta x}{2} \right] \delta t$$

Hence net gain of heat from the two faces is

$$K \ \delta x \delta y \delta z \delta t \ \frac{\partial^2 U}{\partial x^2}$$

We have two similar contributions from the remaining two pairs of parallel faces. Hence total gain of heat in the element is

$$K \ \delta x \delta y \delta z \delta t \left[\frac{\partial^2 U}{\partial x^2} + \frac{\partial^2 U}{\partial y^2} + \frac{\partial^2 U}{\partial z^2} \right]$$

This will increase the temperature U by

$$\frac{\partial U}{\partial t} \ \delta t$$

$$K \ \delta x \delta y \delta z \delta t \left[\frac{\partial^2 U}{\partial x^2} + \frac{\partial^2 U}{\partial y^2} + \frac{\partial^2 U}{\partial z^2} \right] = \rho \delta x \delta y \delta z \ c. \ \frac{\partial U}{\partial t} \ \delta t$$

$$\frac{\partial^2 U}{\partial x^2} + \frac{\partial^2 U}{\partial y^2} + \frac{\partial^2 U}{\partial z^2} = \frac{\rho c}{K} \frac{\partial U}{\partial t}$$

i.e $\quad \nabla^2 U = \frac{\rho c}{K} \frac{\partial U}{\partial t}$

Put

$$\frac{K}{\rho c} = k$$

where k is known as the thermal diffusivity of the material.

$$\therefore \frac{\partial U}{\partial t} = k \nabla^2 U$$

This is known as the heat equation or equation of diffusion. If the flow of heat is uniflow i.e., in one direction say the x - axis then the last equation becomes

$$\frac{\partial U}{\partial t} = k \frac{\partial^2 U}{\partial x^2}$$

When the flow of heat is steady i.e., independent of the time, the heat equation reduces to
$\nabla^2 U = o$

i.e., the temperature U satisfies Laplace's equation.

Example 1

A semi-infinite solid $x \geq o$ is initially at zero temperature. A constant temperature Uo >0 is applied at time t = 0 to the face x = 0 and is maintained. Find the temperature distribution at any point x of the solid and at any instant.

Solution:

Here we have to solve the beat equation

$$\frac{\partial U}{\partial t} = k \frac{\partial^2 U}{\partial x^2} \qquad x > o, t > o$$

subject to the boundary conditions

$$U(x,o) = o, \ U(o,t) = U_o$$

Transforming the differential equation w.r.t. t we get

$$s u(x, s) - U(x,o) = k u_{xx}(x,s)$$

$$\frac{d^2u}{dx^2} - \frac{s}{k}\,u = 0$$

$$\therefore \quad u(x,s) = A e^{\sqrt{s/k}\,x} + B e^{-\sqrt{s/k}\,x}$$

Since U(x,t) is bounded and consequently u(x,s) is also bounded, then A =0 and

$$u(x,s) = B e^{-\sqrt{s/k}\,x}$$

Since $U(o,t) = U_o$ then transforming we get $u(o,s) = \dfrac{U_o}{s}$

$$\therefore \quad \frac{U_o}{s} = B \quad \text{and hence} \quad u(x,s) = \frac{U_o}{s}e^{-\sqrt{s/k}\,x}$$

$$\therefore \quad U(x,t) = U_o\,\mathrm{erfc}\left(\frac{x}{2\sqrt{kt}}\right)$$

Example 2

Same as in problem I but with $U(o,t) = V(t)$.

$$u(o,s) = v(s) \qquad \therefore \quad B = v(s)$$

$$\therefore \quad u(x,s) = v(s)\,e^{-\sqrt{s/k}\,x}$$

$$\text{Now} \quad L^{-1}\left\{e^{-\sqrt{s/k}\,x}\right\} = \frac{x}{2\sqrt{\pi k}}\,t^{-3/2}\,e^{-\frac{x^2}{4kt}}$$

Hence according to a theorem on convolution we have

$$U(x,t) = \frac{x}{2\sqrt{\pi k}}\int_o^t V(t-\lambda)\,\lambda^{-3/2}\,e^{-\frac{x^2}{4k\lambda}}\,d\lambda .$$

Example 3

A solid is bounded by the infinite planes x=o and x=1.

These planes are kept at zero temperature. If at time the distribution of temperature is

U(x,o) $= 4 \sin 2\pi x$,

Find the temperature at any point of the solid and at any instant.

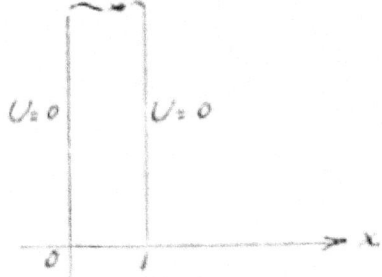

Solution:

$$\frac{\partial U}{\partial t} = k \frac{\partial^2 U}{\partial x^2}$$

$U(o,t) = o. \quad U(1,t) = o, \quad U(x,o) = 4 \sin 2\pi x$

$s\, u(x,s) - U(x,o) = k\, u_{xx}(x,s)$

$$\frac{d^2 u}{dx^2} - \frac{s}{k} u = - \frac{4}{k} \sin 2\pi x$$

$\therefore u(x,s) = A e^{\sqrt{s/k}\, x} + B e^{-\sqrt{s/k}\, x} + \dfrac{4}{s + 4\pi^2 k} \sin 2\pi x$

Now $u(o,s) = o \quad \therefore o = A + B$

$\qquad u(1,s) = o \quad \therefore o = A e^{\sqrt{s/k}} + B e^{-\sqrt{s/k}}$

$\therefore A = B = o$

$$u(x,s) = \frac{4}{s + 4\pi^2 k} \sin 2\pi x$$

$\therefore U(x,t) = 4 e^{-4\pi^2 k t} \sin 2\pi x$

Example 4

A semi-infinite insulated bar is initially at zero temperature. The bar is placed along the positive x -.axis. At t = o a quantity of heat is instantaneously generated at the point x = a where a is positive. Find the temperature distribution along the bar at any instant.

313

Solution:

The partial differential equation satisfied by the temperature U *is*

$$\frac{\partial U}{\partial t} = k \frac{\partial^2 U}{\partial x^2} \qquad (x > 0, t > 0)$$

The boundary conditions are

(i) $U(x,0) = 0$, (ii) $U(x,t)$ is bounded, (iii) $U(a,t) = Q\delta(t)$ where Q is a constant and $\delta(t)$ is the Dirac delta function.

$$su(x,s) - U(x,0) = k\, u_{xx}(x,s)$$

$$\therefore \quad \frac{d^2 u}{dx^2} - \frac{s}{k} u = 0 \qquad \therefore \quad u(x,s) = A e^{\sqrt{s/k}\, x} + B e^{-\sqrt{s/k}\, x}$$

From the boundary condition $A = 0$

$$\therefore \quad u(x,s) = B e^{-\sqrt{s/k}\, x}$$

Transforming $U(a,t) = Q\delta(t)$ we get

$$u(a,s) = Q \qquad \therefore \quad Q = B e^{-\sqrt{s/k}\, a} \qquad \text{i.e} \quad B = Q e^{\sqrt{s/k}\, a}$$

$$u(x,s) = Q e^{\sqrt{s/k}\, a} \cdot e^{-\sqrt{s/k}\, x}$$

i e $$u(x,s) = Q e^{-\sqrt{s/k}\,(x-a)}$$

Now $$L^{-1}\left\{ e^{-a\sqrt{s}} \right\} = \frac{a}{2\sqrt{\pi t^3}} e^{-\frac{a^2}{4t}}, \qquad a > 0$$

$$\therefore U(x,t) = Q \frac{x-a}{2\sqrt{\pi k t^3}} e^{-\frac{(x-a)^2}{4kt}}$$

Example 5

A semi-infinite solid $x \geq 0$ with zero initial temperature has a constant heat flux C applied to the face x=0 so that
$$- KU_x (0,t) = C.$$

Find the temperature at any instant and at any point x of the solid.

Solution:

$$\frac{\partial U}{\partial t} = k \frac{\partial^2 U}{\partial x^2} \ , \quad U(x,0) = 0 \ , \quad - K U_x(0,t) = C$$

$$su(x,s) - U(x,0) = k u_{xx}(x,s)$$

$$\therefore \frac{d^2 u}{dx^2} - \frac{s}{k} u = 0 \quad \therefore \ u(x,s) = Ae^{\sqrt{s/k}\, x} + Be^{-\sqrt{s/k}\, x}$$

From the boundedness condition $A = 0$ $\therefore u(x,s) = Be^{-\sqrt{s/k}\, x}$

Transforming the condition $-KU_x(0,t) = C$ w.r.t. t we get

$$- Ku_x(0,s) = \frac{C}{s} \quad \therefore \ u_x(0,s) = - \frac{C}{Ks}$$

Now $u_x(x,s) = - B\sqrt{s/k}\ e^{-\sqrt{s/k}\, x}$

$$\therefore \quad u_x(0,s) = - B\sqrt{s/k}$$

$$\therefore \ - \frac{C}{Ks} = - B\sqrt{s/k} \quad \text{i.e } B = \frac{C\sqrt{k}}{Ks^{3/2}}$$

$$\therefore \ u(x,s) = \frac{C\sqrt{k}}{Ks^{3/2}} e^{-\sqrt{s/k}\, x}$$

Now $L^{-1}\left\{ s^{-\frac{3}{2}} e^{-a\sqrt{s}} \right\} = 2\sqrt{\frac{t}{\pi}} e^{-\frac{a^2}{4t}}$

$$- a \ \mathrm{erfc}\left(\frac{a}{2\sqrt{t}}\right), \ a > 0$$

$$\therefore \ U(x,t) = \frac{C\sqrt{k}}{K}\left[2\sqrt{\frac{t}{\pi}} e^{-\frac{x^2}{4kt}} - \frac{x}{\sqrt{k}} \mathrm{erfc}\left(\frac{x}{2\sqrt{kt}}\right) \right]$$

i.e $U(x,t) = \frac{C}{K}\left[2\sqrt{\frac{kt}{\pi}} e^{-\frac{x^2}{4kt}} - x\,\mathrm{erfc}\left(\frac{x}{2\sqrt{kt}}\right) \right]$

Example 6

A bar of length l is initially at temperature U_o. The end $x = o$ is insulated while the end $x = l$ is suddenly given the constant temperature U_1. Find the temperature at any point x of the bar and at any instant assuming the surface of the bar to be insulated.

Solution:

$$\frac{\partial U}{\partial t} = k \frac{\partial^2 U}{\partial x^2} \qquad (0 < x < l, t > 0)$$

Boundary conditions are

$$U(x,0) = U_o \quad , \quad U_x(0,t) = 0 \ , \ U(l,t) = U_1$$

$$su(x,s) - U(x,0) = ku_{xx}(x,s)$$

i.e $\quad su(x,s) - U_o = ku_{xx}(x,s)$

$$\frac{d^2 u}{dx^2} - \frac{s}{k} u = - \frac{U_o}{k}$$

$\therefore \quad u(x,s) = A \cosh \sqrt{s/k}\, x + B \sinh \sqrt{s/k}\, x + \frac{U_o}{s}$

Transforming $\quad U_x(0,t) = 0 \quad$ we get $\quad u_x(0,s) = 0$

$\therefore \quad B = 0$ Hence $\quad u(x,s) = A \cosh \sqrt{s/k}\, x + \frac{U_o}{s}$.

Transforming $\quad U(l,t) = U_1 \quad$ we get $\quad u(l,s) = \frac{U_1}{s}$

$\therefore \quad \frac{U_1}{s} = A \cosh \sqrt{s/k}\, l + \frac{U_o}{s} \quad \therefore \ A = \frac{U_1 - U_o}{s \cosh(\sqrt{s/k}\, l)}$

$\therefore \quad u(x,s) = \frac{U_o}{s} + (U_1 - U_0) \frac{\cosh(\sqrt{s/k}x)}{s \cosh(\sqrt{s/k}\, l)}$

$\therefore \quad U(x,t) = U_0 + (U_1 - U_o) L^{-1} \left\{ \frac{\cosh(\sqrt{s/k}\, x)}{s \cosh(\sqrt{s/k}\, l)} \right\}$

According to the theory of the inversion integral we have

$$L^{-1} \left\{ \frac{\cosh(\sqrt{s/k}\, x)}{s \cosh(\sqrt{s/k}\, l)} \right\} = \text{sum of residues of } \frac{e^{st} \cosh(\sqrt{s/k}\, x)}{s \cosh(\sqrt{s/k}\, l)}$$

at its poles. The poles are the zeros of $s \cosh (\sqrt{s/k}\, l)$ i.e $s = 0$ and zeros of $\cosh (\sqrt{s/k}\, l)$

Now consider $\cosh u = 0$ $\therefore$ $\dfrac{e^{u} + e^{-u}}{2} = 0$

$\therefore$ $e^{u} = -e^{-u}$ or $e^{2u} = -1 = e^{\pi i + 2n \pi i}$

where $n = 0, 1, 2, 3, \cdots$

$\therefore$ $2u = (2n + 1) \pi i$ i.e $u = (n + \tfrac{1}{2}) \pi i$

$\therefore$ $\cosh(\sqrt{s/k}\, l) = 0$ if $\sqrt{s/k}\, l = (n + \tfrac{1}{2}) \pi i$

i.e $s = -\dfrac{(2n + 1)^2 \pi^2 k}{4\, l^2}$, $n = 0, 1, 2, 3 \cdots$

which can also be written

$$s = -\frac{(2n - 1)^2 \pi^2 k}{4\, l^2}, \quad n = 1, 2, 3, \cdots$$

Hence zeros of $s \cosh (\sqrt{s/k}\, l)$ are

$$s = 0, \; s = -\frac{(2n - 1)^2 \pi^2 k}{4\, l^2}, \quad n = 1, 2, 3, \ldots$$

Residue at $s = 0$ is $\displaystyle \lim_{s \to 0} s\, \frac{e^{st} \cosh (\sqrt{s/k}\, x)}{s \cosh (\sqrt{s/k}\, l)} = 1$

Residue at $s = -\dfrac{(2n-1)^2 \pi^2 k}{4 l^2} = s_n$ is

$$\lim_{s \to s_n} (s - s_n) \left[\frac{e^{st} \cosh (\sqrt{s/k}\, x)}{s \cosh (\sqrt{s/k}\, l)} \right]$$

$$= \left[\lim_{s \to s_n} \frac{s - s_n}{\cosh (\sqrt{s/k}\, l)} \right] \left[\lim_{s \to s_n} \frac{e^{st} \cosh (\sqrt{s/k}\, x)}{s} \right]$$

$$= \left[\lim_{s \to s_n} \frac{1}{\sinh(\sqrt{s/k}l)\,(l/2\sqrt{ks})} \right] \left[\lim_{s \to s_n} \frac{e^{st}\cosh(\sqrt{s/k}\,x)}{s} \right]$$

$$= \frac{4(-1)^n}{(2n-1)\pi} \, e^{-\frac{(2n-1)^2\pi^2 kt}{4l^2}} \cos\frac{(2n-1)\pi x}{2l}$$

according to L'Hiospital's rule in evaluating indeterminate forms, and the fact that

cosh iu $=$ cos u and sinh iu $=$ i sin u.

$$\therefore \; U(x,t) = U_1 +$$

$$\frac{4(U_1-U_o)}{\pi} \sum_{n=1}^{\infty} \frac{(-1)^n}{2n-1} \, e^{-\frac{(2n-1)^2\pi^2 kt}{4l^2}} \cos\frac{(2n-1)\pi x}{2l}.$$

Example 7

An infinitely long circular cylinder of radius unity is initially at temperature Uo. At time t = 0 , a temperature 0°C is applied to the surface and is maintained. Determine the temperature distribution over the cylinder at any instant t.

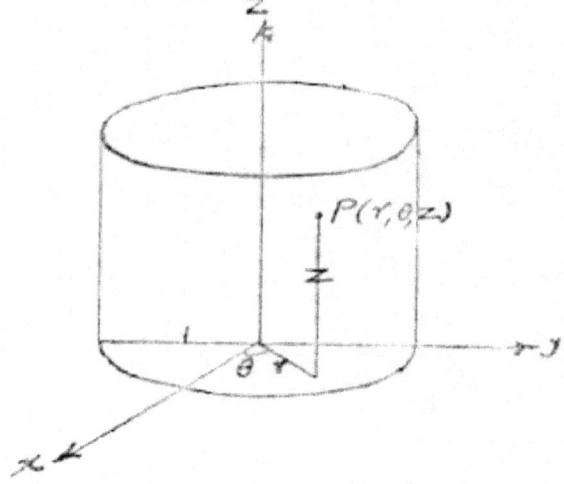

Solution:

The equation satisfied by the temperature U is

$$\frac{\partial U}{\partial t} = k \; \nabla^2 U$$

318

Now in cylindrical coordinates the Laplacian operator takes the form

$$\nabla^2 = \frac{\partial^2}{\partial r^2} + \frac{1}{r}\frac{\partial}{\partial r} + \frac{1}{r^2}\frac{\partial^2}{\partial \theta^2} + \frac{\partial^2}{\partial z^2}$$

and hence the heat equation becomes

$$\frac{\partial U}{\partial t} = k\left[\frac{\partial^2 U}{\partial r^2} + \frac{1}{r}\frac{\partial U}{\partial r} + \frac{1}{r^2}\frac{\partial^2 U}{\partial \theta^2} + \frac{\partial^2 U}{\partial z^2}\right]$$

Since in this problem, due to symmetry, it is independent of θ and z, then

$$\frac{\partial U}{\partial t} = k\left(\frac{\partial^2 U}{\partial r^2} + \frac{1}{r}\frac{\partial U}{\partial r}\right) \qquad 0 < r < 1$$

Now it is more convenient to take this equation in the form

$$\frac{\partial U}{\partial t} = \frac{\partial^2 U}{\partial r^2} + \frac{1}{r}\frac{\partial U}{\partial r}$$

and then in the final result t is changed to kt.

The boundary conditions are
$$U(r,o) = U_o \,, \; U(1,t) = o \text{ and } U \text{ is bounded.}$$

Transform the heat equation w r.t. t
$$su(r,s) - U(r,o) = \frac{d^2 u}{dr^2} + \frac{1}{r}\frac{du}{dr}$$

$$\therefore \; \frac{d^2 u}{dr^2} + \frac{1}{r}\frac{du}{dr} - su = -U_o$$

$$\therefore \; u(r,s) = AJ_o(i\sqrt{s}\,r) + BY_o(i\sqrt{s}\,r) + \frac{U_o}{s}$$

Since u(r,s) is finite then B = o

$$\therefore \; u(r,s) = AJ_o(i\sqrt{s}\,r) + \frac{U_o}{s}$$

Transforming U(1,t) = o we get u(1,s) = o

319

$$\therefore 0 = AJ_0(i\sqrt{s}) + \frac{U_o}{s} \qquad \text{i.e } A = -\frac{U_o}{sJ_0(i\sqrt{s})}$$

$$\therefore n(r,s) = \frac{U_o}{s} - \frac{U_o J_0(i\sqrt{s}\,r)}{sJ_0(i\sqrt{s})}$$

$$\therefore U(r,t) = U_o - U_o\, L^{-1}\left\{\frac{J_0(i\sqrt{s}\,r)}{sJ_0(i\sqrt{s})}\right\}$$

Now according to the inversion integral formula

$$L^{-1}\left\{\frac{J_0(i\sqrt{s}\,r)}{sJ_0(i\sqrt{s})}\right\} = \text{sum of residues of } \frac{e^{st}J_0(i\sqrt{s}\,r)}{sJ_0(i\sqrt{s})} \text{ at its poles.}$$

The poles are zeros of $sJ_0(i\sqrt{s})$ i.e $s=0$ and zeros of $J_0(i\sqrt{s})$.

Now $J_0(i\sqrt{s})$ has simple zeros at $i\sqrt{s} = \lambda_1, \lambda_2, ..., \lambda_n, ...$

i.e $s = -\lambda_{in}^2$ where $n=1,2,3,...$

Residue at $s=0$ is $\displaystyle\lim_{s\to 0} s\, \frac{e^{st}J_0(i\sqrt{s}\,r)}{sJ_0(i\sqrt{s})} = 1$

Residue at $s = -\lambda_n^2$ is

$$\lim_{s\to -\lambda_n^2} (s+\lambda_n^2)\, \frac{e^{st}J_0(i\sqrt{s}\,r)}{sJ_0(i\sqrt{s})}$$

$$= \left[\lim_{s\to -\lambda_n^2} \frac{s+\lambda_n^2}{J_0(i\sqrt{s})}\right]\left[\lim_{s\to -\lambda_n^2} \frac{e^{st}J_0(i\sqrt{s}\,r)}{s}\right]$$

$$= \left[\lim_{s\to -\lambda_n^2} \frac{1}{J_0'(i\sqrt{s})i/2\sqrt{s}}\right]\left[\frac{e^{-\lambda_n^2 t}J_0(\lambda_n r)}{-\lambda_n^2}\right]$$

$$= -\frac{2e^{-\lambda_n^2 t}J_0(\lambda_n r)}{\lambda_n J_1(\lambda_n)}$$

according to L' Hospital's rule in evaluating indeterminate forms and the fact that $J_0'(x) = -J_1(x)$.

320

$$\therefore \quad U(r,t) = U_o - U_o\left[1 - \frac{2\sum\limits_{n=1}^{\infty} e^{-\lambda_n^2 t} J_o(\lambda_n r)}{\lambda_n J_1(\lambda_n)}\right]$$

Changing t into kt we get

$$U(r,t) = 2U_o \sum_{n=1}^{\infty} \frac{e^{-k\lambda_n^2 t} J_0(\lambda_n r)}{\lambda_n J_1(\lambda_n)}$$

Example 8

A semi-infinite solid $x \geq 0$ has zero initial temperature.

The temperature at x=o is given by

$$U(o,t) = U_o \qquad o < t < t_o$$

$$= o \qquad t > t_o$$

Determine the temperature at any point x of the solid and at any instant t.

Solution:

$$\frac{\partial U}{\partial t} = k \frac{\partial^2 U}{\partial x^2}$$

Boundary conditions are

$$U(x,o) = o, \qquad U(o,t) = U_o \qquad o < t < t_o$$

$$= o \qquad t > t_o$$

$$su(x,s) - U(x,o) = ku_{xx}(x,s)$$

$$\frac{d^2 u}{dx^2} - \frac{s}{k} u = o \quad \therefore \quad u(x,s) = Ae^{\sqrt{s/k}\, x} + Be^{-\sqrt{s/k}\, x}$$

Since U is bounded and consequently u, then A = o

$$u(x,s) = Be^{-\sqrt{s/k}\, x}$$

Since $u(o,s) = \frac{U_o}{s}\left(1 - e^{-t_o s}\right)$

$$\therefore \quad B = \frac{U_o}{s}\left(1 - e^{-t_o s}\right)$$

$$u(x,s) = \frac{U_o}{s}\left(1 - e^{-t_o s}\right) e^{-\sqrt{s/k}\, x}$$

i.e $u(x,s) = U_o\left(1 - e^{-t_o s}\right) \cdot \frac{1}{s}\, e^{-\sqrt{s/k}\, x}$

Now $L^{-1}\left\{ \frac{1}{s}\, e^{-a\sqrt{s}} \right\} = \text{erfc}\, \frac{a}{2\sqrt{t}}$, $a > 0$

$\therefore\ U(x,t) = U_o\, \text{erfc}\, \frac{x}{2\sqrt{kt}}$ $\quad o < t < t_o$

$= U_o\, \text{erfc}\, \frac{x}{2\sqrt{kt}} - U_o\, \text{erfc}\, \frac{x}{2\sqrt{k(t - t_o)}}$ $\quad t > t_o$

6.5. Exercise on using Laplace Transformation in solving Linear Partial Differential Equations

(1) Solve the partial differential equation

$$U_{xx}(x,t) + U_{tx}(x,t) - 2U_{tt}(x,t) = o \qquad (x > o, t > o)$$

subject to the following boundary conditions

$U(x,o) = U_t(x,o) = o$, $\lim\limits_{x \to \infty} U(x,t) = o$, $U(o,t) = F(t)$.

[Ans. $\quad U(x,t) = o \qquad o < t < 2x$

$\qquad\qquad = F(t-2x) \qquad t > 2x]$

(2) Solve the following boundary value problem

$Y_x(x,t) + x\, Y_t(x,t) = o$

$Y(x,o) = o,\qquad Y(o,t) = t$

$\Big[$ Ans. $\quad Y(x,t) = o \qquad o < t < \dfrac{x^2}{2}$

$\qquad\qquad = t - \dfrac{x^2}{2} \qquad t > \dfrac{x^2}{2} \Big]$

(3) Solve the partial differential equation

$$Y_{tt}(x,t) = a^2\, Y_{xx}(x.t) \qquad x > o, t > o$$

with the following boundary conditions

$Y(x,o) = o,\ Y_t\ (x,o) = -u_0,\ Y(o,t) = o,\ \lim\limits_{x \to \infty} Y_x(x,t) = o$

$$\left[\ \text{Ans.}\ \ Y(x,t)\ =\ -u_0 t \qquad o < t < \frac{x}{a} \right.$$

$$\left. = -u_0\ \frac{x}{a} \qquad t > \frac{x}{a} \ \right]$$

(4) Find a bounded solution of

$$x\ \frac{\partial U}{\partial x} + \frac{\partial U}{\partial y} = xe^{-y} \qquad o < x < 1,\ y > o$$

which satisfies $U(x,o) = x \qquad o < x < 1$

$[$ Ans. $\ \ U(x,y) = xe^{-y}(1+y)]$

(5) Solve the equation

$$\frac{\partial U}{\partial t} + x\ \frac{\partial U}{\partial x} + U = x \qquad x > o,\ t > o$$

with the boundary conditions
$U(o,t) = o,\ \ U(x,o) = o$

$[$ Ans. $\ \ U(x,t) = \tfrac{1}{2} x\ (1-e^{-2t})\ ]$

(6) Show that *in* solving the partial differential equation
$\dfrac{\partial^2 Y}{\partial t^2} = a^2\ \dfrac{\partial^2 Y}{\partial x^2}$ we can solve the equation $\dfrac{\partial^2 Y}{\partial x^2} = \dfrac{\partial^2 Y}{\partial t^2}$
and then in the result we replace t by at.

(7) Solve the equation

$$\frac{\partial^2 Y}{\partial x^2} = 16\ \frac{\partial^2 Y}{\partial t^2} \qquad x > o,\ t > o$$

subject to the following boundary conditions

$$Y(x,o) = o,\ Y_t\ (x,o) = -1,\ Y(o,t) = t^2,\ \lim\limits_{x \to \infty}\ Y(x,t)\ \text{exists}$$

for fixed $t > o$

[Ans. $Y(x,t) = -t$ $o \leqslant t \leqslant 4x$

          $= (t-4x)^2 - 4x$ $t \geqslant 4x$]

(8) Solve the equation

$$\frac{\partial Y}{\partial x} + 4 \frac{\partial Y}{\partial t} = - 8t \qquad x > o , t > o$$

$Y(x,o) = o, \; Y(o,t) = 2t^2$

[Ans. $Y(x,t) = - t^2$ $o \leqslant t \leqslant 4x$

          $= - t^2 + 3(t-4x)^2$ $t \geqslant 4x$]

(9) Solve the equation

$$\frac{\partial Y}{\partial x} + 2 \frac{\partial Y}{\partial t} = 4t \qquad x > o, t > o$$

subject to the boundary conditions

$Y(x,o) = o, \qquad Y(o,t) = 2t^8$

[Ans. $Y(x,t) = t^2$ $o \leqslant t \leqslant 2x$

          $= t^8 + 2 (t \quad 2x)^8 - (t-2x)^2$ $t \geqslant 2x$]

(10) ---

$$\frac{\partial^2 Y}{\partial t^2} = 4 \frac{\partial^2 Y}{\partial x^2} \qquad x > o, \; t > o$$

$Y(x,o) = o, \; Y_t(x,o) = 2, \; Y(o,t) = \sin t$

$$\lim_{x \to \infty} Y(x,t) \text{ exists for } t > o .$$

[Ans. $Y(x,t) = 2t$ $o \leqslant t \leqslant \frac{x}{2}$

          $= 2t + \sin (t - \frac{x}{2}) - 2 (t - \frac{x}{2})$,

          $t \geqslant \frac{x}{2}$] .